DIFFUSION AND HEAT EXCHANGE
IN CHEMICAL KINETICS

DIFFUSION
AND HEAT EXCHANGE
IN CHEMICAL
KINETICS

By D. A. FRANK-KAMENETSKII

Translated by N. Thon

1955

Princeton University Press

Princeton, New Jersey

This book, originally published in Russian in 1947, was issued through the Publishing House of the U.S.S.R. for the Institute of Chemical Physics of the U.S.S.R. Academy of Sciences; N. N. Semenov, Responsible Editor.

This translation was prepared under contract with the National Science Foundation. Officers, agents, and employees of the United States Government, acting within the scope of their official capacities, are granted an irrevocable, royalty-free, nonexclusive right and license to reproduce, use, and publish or have reproduced, used and published, in the original or other language, for any governmental purposes, all or any portion of the translation.

FOREWORD

In recent years, a new original trend has developed in chemical kinetics, aiming at a complex study of the chemical process in combination with the physical processes of transfer of heat and matter. Unexpectedly, it appeared possible to synthesize such apparently distant branches of science as chemical kinetics, on the one hand, and the theory of heat transfer, diffusion, hydrodynamics, on the other. Processes which earlier, in the classic chemical kinetics, were viewed as perturbations of the course of a chemical reaction, took on a particular interest in combination with the chemical process.

This combination of kinetics with the theory of diffusion, heat transfer, and hydrodynamics yielded a series of new methods of study of rates of reactions, and laid the scientific foundations for the theory of such technically important processes as combustion and dissolution, the basic process of chemical technology.

Investigators of many countries have taken part in the development of this trend in science. Predvoditelev, Knorre, Vulis (in the theory of combustion of coal), Zharovnikov (in the theory of processes and apparatus of chemical technology), Temkin, Boreskov (in the theory of catalysis), in the U.S.S.R.; Hottel, Mayers, Burke and Schumann, Sherwood, Chilton and Colburn, Lewis and von Elbe, Damköhler, Fischbeck, abroad.

At the Institute of Chemical Physics of the Academy of Sciences of the U.S.S.R. which devotes itself especially to problems of kinetics of chemical reactions and the theory of combustion, work has been in progress particularly in macroscopic kinetics. It suffices to quote the well known work of Elovich, Todes, Zeldovich

v

FROM THE AUTHOR

> "I not only saw from other authors, but am convinced by my own art, that chemical experiments combined with physical, show peculiar effects."

> M. V. Lomonosov
> (Project of foundation of a chemical laboratory at the Imperial Academy of Sciences, March, 1745.)

The theme of this book is the role of physical factors in the course of chemical reactions.

Lomonosov, two hundred years ago, foresaw how fruitful the combination of chemistry and physics can be. To this statement, the development of science up to the present day provides ever new confirmation. Many have contributed to the ideas to which this book is devoted.

I express my deep gratitude to my teachers: Ya. B. Zeldovich, L. D. Landau, N. N. Semenov. I cordially thank my closest collaborators: Erna Blyumberg, N. Ya. Buben, Ts. M. Klibanov, I. E. Salnikov, Elena Fridman. For their interest and valuable advice and comments, I am indebted to G. N. Abramovich, A. A. Andronov, A. P. Vanichev, V. V. Voevodskii, L. A. Vulis, V. I. Goldanskii, M. V. Keddysh, G. F. Knorre, E. M. Minskii, M. B. Nieman, V. V. Pomerantsev.

CONTENTS

CHAPTER I: INTRODUCTION

CHAPTER II: DIFFUSIONAL KINETICS

CONTENTS

CHAPTER III: THE STEPHAN FLOW

CHAPTER IV: NONISOTHERMAL DIFFUSION

CHAPTER V: CHEMICAL HYDRODYNAMICS

CONTENTS

CHAPTER VI: THEORY OF COMBUSTION FROM THE POINT
OF VIEW OF THE SIMILITUDE THEORY

CHAPTER VII: TEMPERATURE DISTRIBUTION IN THE REACTION
VESSEL AND STATIONARY THEORY OF
THE THERMAL EXPLOSION

CHAPTER VIII: FLAME PROPAGATION

CHAPTER IX: THERMAL REGIME OF HETEROGENEOUS
EXOTHERMAL REACTIONS

CONTENTS

CHAPTER X: PERIODIC PROCESSES IN CHEMICAL KINETICS

CHAPTER I: INTRODUCTION

Classic chemical kinetics studies the course of a chemical reaction under idealized conditions, namely at a temperature constant in time and space, and at concentrations constant in space.

The task of macroscopic kinetics is to study the chemical reaction under the real conditions of its occurrence in nature or in industry, i.e., with the physical processes attendant on the chemical reaction included. The most important of these physical processes are the diffusion of the reactants and products and the evolution and propagation of heat. Both these processes are strongly influenced by the hydrodynamic conditions, i.e., the kind of motion of the gas or liquid which determines the convective transfer of heat and of matter.

Thus, the specific scope of macroscopic kinetics is the study of the role of diffusion, heat transfer, and convection (i.e., the motion of the gas or liquid) in the course of chemical reactions.

Investigation of the complex processes which, along with the chemical conversion, are governed also by transfer of heat and of matter, has a three-fold purpose.

In the first place, it yields the laws of the course of the chemical process under conditions encountered

1

I. INTRODUCTION

Simple and Complex Reactions

All reactions, whether homogeneous or hetero-geneous, can be classified into simple or complex.

We apply the term simple to reactions in which the rate depends only on the concentrations of the original reactants and does not depend on the concentrations of the reaction products.

By the term complex, we shall designate reactions in which the rate depends on the concentrations of both the reactants and the final products, or, much more frequently, of the intermediate products.

As the concentration of the latter is often difficult to determine, and is subject to considerable variation with time, that dependence will simulate an apparent dependence of the rate of the reaction on time.

Order of the Reaction and Energy of Activation

The dependence of the reaction rate on the concentrations of the participating reactants is usually described by a power law of the form

$$v = kC_A^{n_A} \cdot C_B^{n_B} \ldots \qquad (I, 1)$$

where v is the rate of the reaction, and $C_A, C_B \ldots$ are the concentrations of the reactants A, B ... taking part in the reaction.

The coefficient k which depends only on the temperature is termed the rate constant, and the exponent n, the order of the reaction. There is a distinction between the order of the reaction with respect to an individual reactant (e.g., n_A is the order with respect to reactant A) and the overall order, equal to the sum of the individual orders. In gaseous reactions, the overall order is also referred to as the order with respect to total pressure.

The temperature dependence of the rate constant is given by Arrhenius' law

$$k = ze^{-E/RT} \qquad (I, 2)$$

Here T is the absolute temperature, R, the gas constant, and E and z are constants characteristic of the given chemical reaction. The magnitude E is termed the activation energy and, z, the pre-exponential factor. The activation energy represents the amount of energy which is necessary for the molecule to possess in order to react.

In the case of a simple reaction, the magnitudes C in formula $(I, 1)$ represent the concentrations of the original reactants only. In the case of complex reactions, the formula includes also the concentrations of the re-action products, sometimes the final ones, but much more frequently the intermediates.

Autocatalysis and Intermediate Products

If the rate of a reaction increases with in-creasing concentration of a product formed in that re-action, it is termed autocatalytic. There can be auto-catalysis by a final or by an intermediate product. The species which causes the increase of the rate is called the active species.

The majority of real chemical processes are com-plex and proceed over active intermediates. Reactions be-tween two stable molecules require a large activation energy and, therefore, such reactions are slow. The actually observed reactions usually proceed by a rounda-bout path which circumvents this high energy barrier[10, 11, 12].

In the case of homogeneous reactions, the first stage usually consists in the formation, from the original reactants, of some kind of intermediates. Their nature is not well enough elucidated at this stage, but it is safe to assume that it can be different in different

instances. Doubtlessly, in many cases, free radicals or
free atoms can play the role of such active intermediates.
In other cases, it can be complex and relatively stable
molecules which, however, for various reasons possess a
high degree of reactivity, as organic peroxides.

The active intermediates react with the original
reactants to form the final products. These steps require
relatively low activation energies (particularly when the
active species are free radicals or atoms) and are therefore very fast. In contrast, the primary formation of the
active intermediates from the stable original molecules
does require a high activation energy and cannot therefore
be fast.

In order for the reaction over active intermediates to proceed rapidly, it is necessary that these
intermediates be regenerated in the course of the reaction; in other words, it is necessary that the reaction
between the intermediate and the original reactant yield
not only the stable final product but also new active
intermediates.

Chain Reactions

Reactions in which active intermediates are regenerated are termed chain reactions. It has been established by Semenov[10], Hinshelwood, and others, that the
majority of the actual homogeneous complex reactions are
chain processes.

There are two characteristic ways in which a
chain reaction can progress; it can be either stationary
or non-stationary. Let x designate the concentration
of the active intermediate. Its change with time is
governed by the kinetic equation

$$\frac{dx}{dt} = n_o + fx - gx \qquad\qquad (I, 3)$$

where n_o, f, g are kinetic constants usually termed as
follows: n_o is the rate of generation of chains, f is
the rate of chain branching, and g is the rate of chain
termination.

By chain generation, one means the initial process
of formation of the active intermediate from the original
reactant; chain branching is the process wherein a molecule
of the active intermediate, reacting with an original reactant, gives rise to two or several molecules of an
active intermediate: chain termination is the process in
which the active intermediate is irrevocably annihilated.
In addition to these processes, included in equation
(I, 3), the chain can be carried on through a process in
which a molecule of the intermediate, reacting with an
original reactant, forms the final product and one new
active-intermediate molecule. While this process does
regenerate the chain, its rate does not figure in equation (I, 3) as it does not result in any change of the
amount of the active intermediate; it merely regenerates
the exact amount that has been expended. But the rate of
chain continuation, multiplied by the concentration of the
active species, determines the rate of the overall conversion of the original reactants into the final products.

Stationary and Nonstationary Course of the Reaction

Solutions of the equation (I, 3) vary depending
on the relative magnitudes of f and g. if g > f, the
course of the reaction is stationary. the concentration
x of the active intermediate will, with time, tend to the
stationary value

$$X = \frac{n_o}{g - f} \qquad (I, 4)$$

Once this stationary value has been attained, the concentration of the active intermediate will remain constant
and the reaction will proceed at the constant rate

I. INTRODUCTION

$$v = kX \qquad\qquad (I, 5)$$

where k is the rate constant for the chain continuation.

Actually, n_0, g, and f depend on the concentrations of the original species and therefore change slowly over the course of the reaction. At the same time, the magnitude X also changes according to equation (I, 4). It is therefore more correct to refer to it not as a stationary but as a quasi-stationary concentration of the active species. The initial change of x, until it reaches its quasi-stationary value X, takes place within a very short time interval during which the concentrations of the original substances have not had time to change significantly.

An entirely different situation will arise if f > g, the course of the reaction will become nonstationary. The solution of equation (I, 3) has the form

$$x = \frac{n_0}{\varphi}(e^{\varphi t} - 1) \qquad\qquad (I, 6)$$

where $\varphi = f - g$, and t is the time. At stages corresponding to t much greater than $\frac{1}{\varphi}$, the concentration of the active species and the rate of the reaction will increase with time proportionally to $e^{\varphi t}$, according to an exponential law. The initial period during which the concentration of the active species and the rate of the reaction are comparatively small is called the induction period. Its duration is of the order of $\frac{1}{\varphi}$.

If the reaction is conducted in such a way that g and f vary, a sharp transition from stationary to nonstationary will occur at g = f. This penomenon is referred to as chain ignition.

Thus far, we have considered the simplest case where only one active species is involved in the reaction and where the kinetic equations are linear. In reality, the equations can be of more complex form; however, in principle, the picture remains the same.

Let the reaction involve several intermediate active species at the concentrations x_i. For the change of any one of them with time, we can write

$$\frac{dx_i}{dt} = F_i(x, a) \qquad (I, 7)$$

where a is the initial concentration, and F some function. Equating the right-hand members to zero, we get the system of algebraic equations.

$$F_1(x, A) = 0; \; F_2(x, A) = 0; \; \ldots \; F_i(x, a) = 0 \quad (I, 8)$$

Solution of this system should give the quasi-stationary concentrations X of the intermediate species as a function of the initial concentrations a

$$X = f(a) \qquad (I, 9)$$

If the system of equations (I, 8) has finite real positive roots, and the conditions derived in our paper[11] are ful-filled, the reaction will take place under stationary con-ditions. After the lapse of a short initial period, the concentrations of all intermediate species will be very close to the quasi-stationary values X and from then on will vary only with the variation of the initial concen-trations. If there are not finite real positive solutions to the system (I, 8), the course of the reaction will be nonstationary, i.e., the concentrations of the active intermediates and the rate of the reaction will increase with time. At constant initial concentrations, this growth will be unlimited, and come to a halt only as a result of exhaustion of reactants.

Mathematically, the transition from the station-ary to the nonstationary course of the reaction corresponds

to a point where the system of equations (I, 8) ceases to
have finite real positive roots. That is the condition
of chain ignition in its most general form.

Heterogeneous Reactions

In heterogeneous reactions, the role of active
intermediates is usually taken over by molecules bound to
the surface by chemical forces, or, following the usual
terminology, chemisorbed to the surface. Therefore, in
the chemical mechanism of a heterogeneous reaction, some
adsorption stage usually plays a decisive role.

The kinetics of actual heterogeneous reactions
is usually strongly complicated by inhomogeneity of the
surface. A real solid surface is never uniform in the
kinetic and energetic sense. Different portions of the
surface are characterized by different values of the heat
of adsorption and the activation energy.

Adsorption on a uniform surface is governed by
Langmuir's adsorption isotherm which expresses the dependence of the amount g of adsorbed substance on its
partial pressure p in the gas phase by the formula

$$g = g_0 \frac{p}{p + b}$$

where g_0 and b are constants.

In the so-called Langmuir kinetics on a uniform
surface, the dependence of the rate of reaction on the
concentrations of the reactants in the gas phase should
be of an analogous form.

On a nonuniform surface, the dependence of the
adsorbed amount of a substance on its partial pressure in
the gas phase is usually described by the Freundlich
isotherm

$$g = Cp^{\frac{1}{n}}$$

where C is a constant, and n > 1.

With this type of adsorption isotherm, the reaction kinetics usually follows a fractional order.

Theoretical analysis of the kinetics of complex
heterogeneous reactions on a nonuniform surface is beyond
the scope of this book. References to the pertinent literature can be found in our review[13]. For a theoretical
interpretation of the Freundlich isotherm, see the work
of Zeldovich[14]

NOTIONS OF DIFFUSION AND HEAT TRANSFER THEORY

Similarity of Diffusion and Heat Transfer Processes

Phenomena of transfer of heat and of matter are
similar. Heat transfer by molecular heat conduction has
its analog in molecular diffusion, and heat transfer by
convection, in convective diffusion. Only heat transfer
by radiation has no analog in transfer of matter.

Thanks to the similarity of the processes of
transfer of heat and of matter, it is not necessary to
consider each class separately. All theoretical and experimental results obtained in the study of heat transfer[15, 16] can be directly applied to diffusion phenomena,
and vice versa.

Heat Conductance and Diffusion in an Immobile Medium

The fundamental law of transfer of heat in an
immobile medium (molecular heat conduction) is Fourier's
law according to which the heat flow q, i.e., the
quantity of heat transferred across unit surface area per
unit time, is proportional to the temperature gradient

$$q = - \lambda \frac{dT}{dx} \qquad (I, 10)$$

where x is the coordinate perpendicular to the surface
across which the heat is transferred, and λ is the heat
conductivity, a physical constant of the substance in
which the heat is propagated. The minus sign indicates
that the heat is transferred in the direction along which
the temperature decreases, i.e., in the direction of the
negative temperature gradient.

The corresponding law for diffusion is Fick's
law according to which the diffusion flow is proportional
to the concentration gradient

$$q = - D \frac{dC}{dx} \qquad\qquad (I,\ 11)$$

where C is the concentration of the diffusing substance,
and q is the diffusion flow, i.e., the quantity of
substance transferred across unit surface area per unit
time. We use the same symbol q for both the flow of
heat and of matter, because of the similarity of heat and
diffusion flow; this can hardly give rise to any misunder-
standings. The proportionality coefficient D is termed
the diffusion constant.

Heat conductance and diffusion in an immobile
medium can be observed in a pure form only in solids. In
liquids and in gases they are unavoidably accompanied by
a motion of the liquid or the gas in free or forced
convention.

Free and Forced Convection

In contrast to molecular heat conduction and
diffusion, where the transfer of heat or of matter is
brought about by the motion of individual molecules, the
transfer of heat or of matter due to a motion of the gas
or the liquid as a whole is termed convection. If that
motion is caused by the same difference of temperatures
or concentrations which determines the transfer of heat
or of matter, the convection is referred to as free or

natural. If the motion is brought about by external forces,
the process is termed forced convection.

In the presence of convection, the laws of
Fourier and of Fick must be supplemented by additional
terms expressing the transfer of heat or of matter by
the mass flow.

With the latter designated by v, Fick's law
takes on the form

$$q = - D \frac{dC}{dx} + v_x C \qquad (I, 11a)$$

and Fourier's law becomes

$$q = - \lambda \frac{dT}{dx} + c\rho v_x T \qquad (I, 10a)$$

where v_x is the component of v in the direction of the
x coordinate, c is the heat capacity, and ρ the
density.

Laminar and Turbulent Flow

The character of the convective transfer of heat
or of matter depends on the type of motion of the gas or
liquid. Depending on hydrodynamic conditions, this motion
can be either laminar or turbulent. A laminar flow is an
ordered stationary motion wherein the velocity at each
point is constant in time, and the velocities at neighbor-
ing points are parallel to each other. A turbulent flow
is a disordered nonstationary motion wherein the velocity
at any point changes continually, with time in a random
fashion.

In a laminar flow, the mechanism of the transfer
of heat or of matter is essentially the same as in an
immobile medium. As before, the transfer takes place
through molecular heat conductance or diffusion, and only
the external conditions vary as a result of the mass flow.

I. INTRODUCTION

Not so in a turbulent flow where the transfer
takes place through turbulent pulsations, i.e., disordered
motions of small volumes of gas or liquid.

Transfer Coefficients

In addition to the similitude, there is one more
deep analogy between the mechanism of the processes of
diffusion and heat transfer, and the process of transfer
of momentum which underlies the resistance of a gas or
liquid to motion. In the absence of turbulence, the intensity of all these three processes is characterized by
the coefficients of molecular transfer. These coefficients will be defined in the following way.

For the transfer of heat, we shall introduce
the so-called thermometric conductivity coefficient a,
related to the usual heat conductivity coefficient λ by

$$a = \frac{\lambda}{c\rho} \qquad\qquad (I, 12)$$

where c is the heat capacity, and ρ the density. In
other words, the temperature conductivity coefficient is
the heat conductivity divided by the heat capacity of unit
volume.

For the transfer of matter, we shall use the
ordinary diffusion coefficient D defined by formula
(I, 11).

For the transfer of momentum, we introduce the
kinematic viscosity υ, related to the usual viscosity μ
by

$$\upsilon = \frac{\mu}{\rho} \qquad\qquad (I, 13)$$

All three transfer coefficients, a, D, and υ have the
same dimension, cm^2/sec. In gases, where the mechanism
of transfer of all three magnitudes, heat, matter, and

momentum is the same and is determined by the thermal
motion of the molecules, all three coefficients, a, D,
and υ are numerically of the same order of magnitude.
In liquids, the kinematic viscosity can be much greater
than the temperature conductivity, and, particularly, the
diffusion coefficient.

According to the kinetic gas theory, the transfer coefficients for an ideal gas are of the order of the
product of the mean free path Λ and the root mean square
velocity U of the molecules

$$a \approx D \approx \upsilon \approx \Lambda U \qquad\qquad (I, 14)$$

Turbulent Exchange Coefficient

In the case of turbulent motion in a gas or
liquid, the role of all three transfer coefficients is
taken over by the so-called turbulent exchange coefficient
A. The disordered turbulent motion of a gas or liquid is
similar to the disordered thermal motion of molecules, except that it involves not individual molecules, but small
volumes of gas or liquid which keep their individuality
for some time. Such small volumes will be referred to as
gaseous or liquid particles. An illustration is Figure 1
which represents the pulsations of the velocities of wind.
The longitudinal dimensions which characterize the turbulent motion are referred to as the scale of the turbulence.

In hydrodynamics, two different methods are used
to describe the motion of a gas or liquid. The first,
proposed by Lagrange, follows the motion of a given individual liquid particle; the second, due to Euler, considers the distribution of velocities in space at a given
moment. Correspondingly, the two basic scales of turbulence are termed, respectively, Lagrangian or Eulerian.
The Lagrangian scale of turbulence is the path along which
the particle keeps its individuality. The Eulerian scale
of turbulence is the mean dimension of such an individual

particle.

One can define a magnitude which, for turbulence, plays the role of a mean free path. That magnitude is the mixing length l. It is related with the turbulence scales, and the form of that relation can depend on the character of the motion.

In the simplest case of isotropic turbulence, when the pulsations of the velocity are the same in all directions, the mixing path coincides with the Lagrangian turbulence scale.

The analog in turbulence of the root mean square velocity of molecular motion is the mean pulsation velocity u.

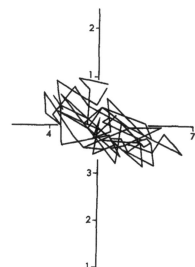

FIGURE 1. PULSATIONS OF WIND VELOCITIES
The abscissa represents the horizontal and the ordinate the vertical components of the velocities, as measured at the airport of Akron, Ohio (USA). The figure is taken from "Problems of turbulence", edited by Velikanov and Shveikovskii (ONTI, Moscow 1936).

In analogy with the gas kinetic theory, the coefficient of turbulent exchange is represented as the product

$$A = lu \qquad (I, 15)$$

Heat Exchange Coefficient

Processes of transfer of heat and matter in convective motion are not always susceptible to analytic calculation, particularly when the motion is turbulent. Therefore, such problems are usually approached with the aid of empirical coefficients.

For heat transfer, the ratio of the heat flow q and the temperature difference ΔT is usually termed the heat exchange coefficient α,

$$q = \alpha \Delta T \qquad\qquad (I, 16)$$

This relation is known as Newton's law of heat exchange. Actually, it is not the expression of a law of nature but merely a definition of the heat exchange coefficient.

Formula (I, 16) does not, of course, solve the problem of calculation of the heat transfer process, but reduces it to the determination of the heat exchange coefficient. This can be done either from experimental data and empirical formulas derived therefrom, or, with the aid of the theory of similitude as will be described below.

Nonetheless, the use of the heat exchange coefficient has proved a very convenient method of calculation and has taken root in practice.

Diffusion Velocity Constant

In analogy with the heat exchange coefficient, we shall introduce, for the transfer of matter in the presence of convection, a magnitude to be termed the diffusion velocity constant, β, defined as the ratio of the diffusion flow q and the concentration difference ΔC,

$$q = \beta \Delta C \qquad\qquad (I, 17)$$

This relation is to be viewed merely as a definition of the diffusion velocity constant β. Inasmuch as the diffusion flow is expressed in moles/cm^2 × sec. and the concentration in moles/cm^3, the dimension of the diffusion velocity constant is cm/sec, i.e., that of a velocity.

Dimensionally, β does not correspond to α, but to $\frac{\alpha}{c\rho}$, where c is the heat capacity and ρ, the density. That is because, in diffusion, there is no magnitude analogous to the heat capacity. Concentration is defined, simply, as the amount of matter per unit volume, whereas temperature is not simply equal to the amount of

I. INTRODUCTION

heat contained in unit volume but to its quotient by $c\rho$, the heat capacity of unit volume.

Similitude Theory

It is seldom possible to find the value of the heat exchange coefficient α or of the diffusion velocity constant β by analytical calculation. In the majority of cases, it is necessary to make use of experimental data. Methods of generalization of the experimental data thus become very important. In this way, experimental results on heat transfer can be utilized for the calculation of diffusion processes and vice versa. Results of experiments made on a small model can be utilized for the calculation of processes in a large-scale setup, and results of experiments made with one substance can be utilized for the calculation of processes with another substance participating.

Ways of such generalizations of experimental data are supplied by the similitude theory. This theory rests on the postulate that the real laws of nature cannot depend on the choice of the system of units of measurement. Therefore, any real law can be represented in the form of a relation between dimensionless magnitudes, the so-called similitude criteria. Once this relation has been established, for certain geometric and physical conditions, by means of an evaluation of experimental data, it can further serve for the calculation of any process taking place under the same geometric and physical conditions but with different dimensions, velocities, and physical properties of the substances. Moreover, the relation between dimensionless magnitudes, obtained through an analysis of heat transfer phenomena, can be directly utilized for the calculation of diffusion processes.

We are interested in the determination of the heat exchange coefficient α and of the diffusion velocity constant β. The similitude theory shows that it is

possible to construct two different dimensionless parameters with these magnitudes. One is the so-called
Nusselt criterion, the other the Margoulis criterion.*

The Nusselt criterion for heat transfer is defined by

$$Nu = \frac{\alpha d}{\lambda} \qquad\qquad (I, 18)$$

where α is the heat exchange coefficient, d, a linear
dimension, and λ the heat conductivity of the medium in
which the heat transfer takes place.

For the diffusion process, Nusselt's criterion
is of the form

$$Nu = \frac{\beta d}{D} \qquad\qquad (I, 19)$$

where α is the diffusion velocity constant, d, a linear
dimension, and D the diffusion coefficient.

The Nusselt criterion has proved to be most convenient for the calculation of transfer processes in an
immobile medium or in a laminar flow. In the instance of
purely molecular transfer, Nusselt's criterion is a
constant magnitude, depending only on the geometric shape
of the body. In the presence of significant turbulence,
another dimensionless magnitude, called the Margoulis
criterion[17], has proved to be more convenient. This magnitude characterizes the ratio of the velocity of transverse transfer of heat or matter and the linear velocity
V of the flow. For the heat transfer process, Margoulis
criterion is expressed by

$$M = \frac{\alpha}{c_\rho V} \qquad\qquad (I, 20)$$

For diffusion, Margoulis criterion is defined simply as
the ratio of the diffusion velocity constant β and the
linear velocity of the flow V,

---

* also called Stanton Group.

I. INTRODUCTION

$$M = \frac{\beta}{V} \qquad\qquad (I, 21)$$

In the limiting case of highly developed turbulence,
Margoulis criterion ought to tend to a constant value.
This limiting case is referred to as the region of pure
turbulence. It cannot be actually reached but only be
approached asymptotically. The conclusion that Margoulis
criterion should tend to constancy at the limit of pure
turbulence follows from the consideration that under these
conditions the process should become independent of
molecular constants and be governed only by magnitudes
characteristic for turbulent transfer.

The similitude theory leads to the conclusion
that for a given geometric shape the Nusselt and the
Margoulis criteria should be functions of other dimensionless magnitudes expressing the physical properties of the
medium and the nature of the motion of the gas or liquid.

The physical properties of the medium in which
the transfer of heat or of matter takes place are characterized by the value of a dimensionless magnitude termed
the Prandtl criterion and are defined as the ratio of two
coefficients of molecular transfer.

For the heat transfer process, Prandtl's criterion is expressed by

$$Pr = \frac{\upsilon}{a} \qquad\qquad (I, 22)$$

and for diffusion, by

$$Pr = \frac{\upsilon}{D} \qquad\qquad (I, 23)$$

where υ is the kinematic viscosity, a, the temperature
conductivity coefficient, and D the diffusion coefficient.

From what has been said above about the order of
magnitude of the coefficients of molecular transfer, the

values of Prandtl's criterion for gases are close to
unity. For liquids, they are much greater than unity.
For viscous liquids, the thermal Prandtl criterion can
attain several hundreds. The diffusional Prandtl criterion is even greater and attains the order of a thou-
sand for ordinary aqueous solutions.

The character of the motion of a gas or liquid
in forced convection is determined by the value of
Reynolds' criterion

$$Re = \frac{Vd}{\upsilon} \qquad\qquad (I, 24)$$

where V is the linear velocity of the flow, d, a linear
dimension, and υ the kinematic viscosity. Reynolds'
criterion is a magnitude of purely hydrodynamic nature.
It includes no magnitudes characteristic of the process of
transfer of heat or matter itself. Therefore, there is
no difference in the definition of the Reynolds' criterion
for the heat transfer or the diffusion process. The value
of Reynolds' criterion determines the nature of the motion
of the gas or liquid in the flow. At low values of the
Reynolds number, the motion is laminar and at high values,
turbulent.

It is shown in the similitude theory that in processes of forced convection the Nusselt or the Margulis
criterion, for given geometric and physical conditions, is
a definite function of the criteria of Reynolds and
Prandtl:

$$Nu = f(Re, Pr) \qquad\qquad (I, 25)$$
$$M = \varphi(Re, Pr) \qquad\qquad (I, 26)$$

The form of that dependence is found by analysis
of experimental data of heat transfer or diffusion. Once
that dependence has been established, it can be used to
calculate the coefficient of heat transfer or the diffusion
velocity constant for any processes taking place under

similar geometric and physical conditions. The only
difference between the calculation of heat transfer and
of diffusion is that in the first instance one has to introduce in formulas (I, 25) and (I, 26) the value of the
thermal, and in the second instance, the value of the
diffusional Prandtl criterion.

Having calculated the value of the Nusselt or
the value of the Margoulis criterion from (I, 25) or
(I, 26), we can easily determine the coefficient of heat
exchange α or the diffusion velocity constant β with
the aid of formulas (I, 18) and (I, 19), or (I, 20) and
(I, 21) which for this purpose are of the form

$$\alpha = \frac{Nu\ \lambda}{d} = Mc\rho V$$

$$\beta = \frac{Nu\ D}{d} = MV$$

The form of the relations (I, 25) or (I, 26) is
different for laminar and for turbulent flow. In laminar
flow, the Nusselt criterion tends to become constant,
whereas in turbulent flow the same applies to Margoulis
criterion. Consequently, (I, 25) is more suitable for
laminar flow and (I, 26), for turbulent flow.

In all transfer processes, one can observe two
limiting regions. When the Reynolds criterion tends to
zero, one observes purely molecular transfer, realized in
an immobile medium or in a laminar flow. In this case the
laws of transfer of heat or matter have the form

$$Nu = const$$

The contrary limiting case corresponds to the Reynolds'
criterion tending to infinity, when the laws of heat
transfer and diffusion tend to the limiting form characteristic of the turbulent region

$$M = const$$

The latter region cannot be actually realized. With the
Reynolds criterion tending to infinity, the heat transfer
and diffusion laws tend to their limiting turbulent form
only logarithmically.

For processes of free convection, the criterion
governing the character of the motion is, instead of
Reynolds' criterion, another dimensionless magnitude called the Grashof criterion and defined by

$$Gr = \frac{gd^3}{v^2} \gamma \Delta T \qquad\qquad (I, 27)$$

where g is the gravity acceleration, d, a linear
dimension, v the kinematic viscosity, γ the volume expansion coefficient of the medium, and ΔT the temperature difference giving rise to the convection.

For gases $\gamma = \frac{1}{T}$, where T is the absolute
temperature, and the Grashof criterion takes the form

$$Gr = \frac{gd^3}{v^2} \cdot \frac{\Delta T}{T} \qquad\qquad (I, 27a)$$

For free convection, the Nusselt or the Margoulis
criterion is a definite function of the Grashof and
Prandtl criteria.

Effective Film

In describing processes for transfer of heat or
matter between a stream of gas or liquid and a solid surface, it is often convenient to introduce the conventional
concept of the effective film. Let us assume that at a
distance from the surface the temperature and the concentration are constant (this assumption is not all too far
from reality in a turbulent flow) and that changes of
these magnitudes occur only in a layer of thickness δ
immediately adjacent to the surface. This fictitious
layer is what is called the effective film. Its thickness

δ is so chosen that the actual intensity of the transfer
is obtained on the assumption that its mechanism is purely
molecular within the film. In this way we have

$$\alpha = \frac{\lambda}{\delta}$$

$$\beta = \frac{D}{\delta}$$

These relations should be considered as definitions of δ.

Comparing (I, 18) and (I, 19), we find

$$\delta = \frac{d}{Nu} \qquad\qquad (I, 28)$$

In this way, the thickness δ of the effective film
appears as an auxiliary magnitude replacing Nusselt's criterion. In diffusion processes, the effective film is
often termed the diffusion layer.

The External and Internal Problem

The influence of Reynolds' criterion on the character of the motion of the liquid differs depending on the
geometric conditions. In hydrodynamics, one distinguishes
two types of problems depending upon the character of the
motion of a liquid - the so-called external problem and
the internal problem.

The external problem is that of a flow around
an isolated body, the overall dimensions of which can be
considered infinite. The linear dimension figuring in
the similitude criteria is in that case the dimension of
the body placed in the flow.

The internal problem deals with a flow within a
tube or channel. In this case, the linear dimension d
is the diameter of the tube.

In the external problem, the transition from
laminar to turbulent takes place without discontinuity.

With gradual change of the Reynolds criterion, all magnitudes characteristic of the flow, in particular the Nusselt and Margoulis criteria, change continuously.

In the internal problem, the transition from laminar to turbulent occurs discontinuously at a certain critical value of the Reynolds criterion, which, for a straight circular tube, lies between 2100 and 2300. This phenomenon is referred to as the hydrodynamic critical condition. At this condition, all magnitudes characteristic of the flow, in particular the Nusselt and Margoulis criteria, change discontinuously.

In the external problem, the purely laminar and the purely turbulent as well are only limiting cases for very small and very large values of the Reynolds criterion.

A characteristic and significant property of the internal problem is the range of purely laminar motion of the liquid, the so-called Poiseuille flow, in which turbulence is entirely absent. This is the only possible flow at values of the Reynolds criterion below the critical.

In this flow, the velocity distribution over the cross-section of the tube satisfies the parabolic law of Poiseuille

$$v = v_0 \left(1 - \frac{r^2}{R^2} \right)$$

Such a distribution is called the laminar velocity profile. In a turbulent flow the velocity profile is considerably steeper. Over the major part of the cross-section, the velocity is almost constant; there is a sharp fall of the velocity only in close vicinity to the wall, in the so-called boundary layer.

Resistance Coefficient and Reynolds' Analogy

In every contact between flowing gas or liquid and a solid body, the flow exerts on the body a definite

I. INTRODUCTION

$$r' = \frac{d}{4} \qquad\qquad (I, 34)$$

Thus, for a circular tube,

$$f = \frac{\zeta}{4} \qquad\qquad (I, 35)$$

The resistance force is composed of two parts, the friction resistance and the form resistance. The latter (or, as it is also called, the pressure resistance) is linked with phenomena of flow separation and formation of a zone of reverse circulation behind the body. The character of these phenomena is illustrated by Figure 2, representing the pattern of the motion in a transverse flow around a cylinder (after data of Lohrisch). A change of the conditions of the rupture of streams in connection with the formation of turbulence in the boundary layer can, under certain circumstances, result in a sharp change of the resistance coefficient in an external flow. This is clearly seen from the graph (Figure 3) giving the dependence of the resistance coefficient on Reynolds' criterion in a flow around a sphere. There is a sharp drop of the resistance coefficient at a Reynolds number of about 100,000. All these phenomena

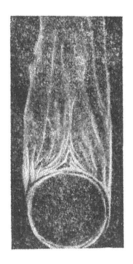

FIGURE 2. Pattern of the motion in transverse flow around a cylinder (after Lohrisch).

are essential only in the external problem, in particular for transverse flow around bodies. For a straight tube with smooth walls, only the friction resistance plays a role.

In the mechanism of internal friction consisting in transfer of momentum, the similitude of diffusion and heat transfer can be extended to frictional resistance. This similitude between heat exchange (and, consequently, also diffusion) and frictional resistance was first established by Reynolds and is called the Reynolds analogy. If the resistance is due only to friction, the resistance coefficient corresponds to the Margoulis criterion, and the two magnitudes are related by the simple expression

$$M = \frac{\zeta}{8} = \frac{f}{2} \qquad\qquad (I, 36)$$

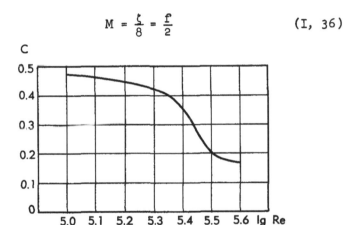

FIGURE 3. Dependence of the resistance coefficient on Reynolds' criterion in a flow around a sphere.

Ordinate: resistance coefficient
Abscissa: log Re

Relations Between the Criteria

The form of the relations between the Nusselt or Margoulis criteria and the Reynolds and Prandtl criteria for systems of a definite geometric shape is determined by an analysis and a generalization of experimental data, or, in simpler cases where there is no influence of turbulence, by analytical calculation.

In the internal problem, we have two entirely different relations for laminar and for turbulent flow conditions. In a flow, the Nusselt criterion depends very little on the Reynolds criterion; in turbulent flow,

formulas are usable or else a Nusselt solution in the
form of an infinite series - very inconvenient for
practical computations.

        The dependence of the mean and the local values
of Nusselt's criterion on Pe $\frac{d}{L}$ and Pe $\frac{d}{z}$ is represented graphically in Figures 4 and 5, taken from the work of
Ya. M. Rubinshtein[19]. The sloping straight line was
traced with the aid of Lévêque's formula, the curve, by
Nusselt's solution; the horizontal line shows the limiting value of Nusselt's criterion, with the corresponding
experimental data of Rubinshtein.

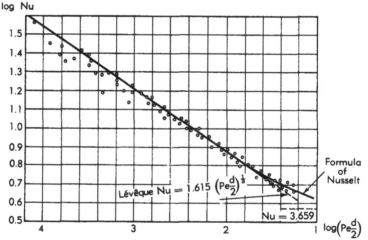

FIGURE 4. MEAN VALUE OF NUSSELT'S CRITERION IN A LAMINAR FLOW,
                FOLLOWING RUBINSHTEIN
The ordinate is log Nu and the abscissa log Pe $\frac{d}{L}$. The inclined line
represents the theoretical solution of Nusselt. Its upper rectilinear
part coincides with Lévêque's formula (I, 30); its applicability is
shown by the dotted sloping straight line. The horizontal dotted line
shows the limiting value of Nusselt's criterion for an infinitely long
tube (I, 38). The circles represent the experimental data of Rubinshtein.

        If in the internal problem, the dependence between
the criteria can be expressed by general formulas, in the
external problem it depends strongly on the shape of the
body and the characteristics of the flow (e.g., the degree
of its turbulization). Ordinarily, Nusselt's criterion
appears to be proportional to Reynolds' criterion to the
power 0.5, this relation holding in a wide range of variation of the Reynolds criterion. The sharp transition

from laminar to turbulent flow is absent in the external problem.

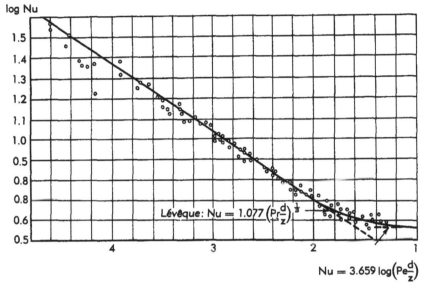

FIGURE 5. LOCAL VALUE OF NUSSELT'S CRITERION IN A LAMINAR FLOW, FOLLOWING RUBINSHTEIN[19]
(The designations are the same as in Figure 4.)

Thus, for the external problem, one can adopt

$$Nu = kRe^m Pr^n \qquad (I, 41)$$

where the exponent m varies between 0.4 and 0.67, and n between 0.3 and 0.4. Both these magnitudes and the coefficients k, depend on the geometric shape of the body and the degree of turbulization of the oncoming flow. The most frequently occurring values of the exponents in formula (I, 41) are

$$m = \frac{1}{2}; \quad n = \frac{1}{3} \qquad (I, 42)$$

In the simplest case of the external problem — longitudinal flow around a plate — these values of the exponents were obtained analytically by Pohlhausen[20].

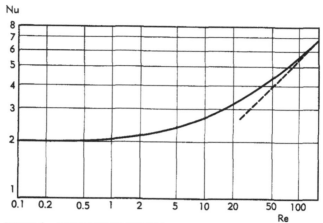

FIGURE 6. DEPENDENCE OF NUSSELT'S CRITERION ON REYNOLDS' CRITERION
IN A FLOW AROUND A SPHERE, ACCORDING TO VYRUBOV AND SOKOLSKII
The Ordinate is Nu and the abscissa Re. The full line corresponds
to the Sokolskii formula (I, 45), the dotted line to the Vyrubov
formula (I, 45). The graph is plotted in a logarithmic scale.

For another simple case, gas flow around a sphere
(i.e., at a value of Prandtl's criterion close to unity),
experimental data obtained by Vyrubov[21] lead, for larger
values of Re, to the formula

$$Nu = 0.54 \sqrt{Re} \qquad (I, 43)$$

This formula holds at Re > 200. At small Re, Nusselt's
criterion for a sphere tends to the limiting constant
value

$$Nu_o = 2 \qquad (I, 44)$$

which is very easily obtained analytically. In the intermediate range at Re < 200, one has the formula of
Sokolskii[22]

$$Nu = 2 \left(1 + 0.08 Re^{2/3} \right) \qquad (I, 45)$$

The complete form of the dependence of Nusselt's criterion
on Reynolds' criterion in the case of a gas flow around
a sphere, according to the experiments of Sokolskii, is

represented in Figure 6. The dotted straight line corresponds to the formula of Vyrubov (I, 43). For thin fibers,
the experiment gave, at small Re, the limiting constant
value of Nusselt's criterion.

$$Nu_0 = 0.45 \qquad\qquad (I, 46)$$

even though the analytical solution of the thermal or
diffusional problem for a cylinder did not yield such a
limit.

In the foregoing, the relations between the criteria were given in the form (I, 25), i.e., the expressions
of Nusselt's criterion were given as is usually done in
the literature. If Nusselt's criterion is known, it is
very easy to compute the criterion of Margoulis, with the
aid of the obvious relation

$$M = \frac{Nu}{RePr} = \frac{Nu}{Pe} \qquad\qquad (I, 47)$$

In this way one can pass from a relation in the form (I, 25)
to the form (I, 26).

The dependence of the resistance coefficient on
Reynolds' criterion is approximately the same as the dependence of Margoulis criterion, as can also be inferred
from Reynolds' analogy [formula (I, 36)].

For turbulent motion in the internal problem,
Blasius' formula

$$f = \frac{0.3164}{\sqrt[4]{Re}} \qquad\qquad (I, 48)$$

is very popular. According to it, the resistance coefficient is inversely proportional to Reynolds' criterion to
the power 0.25, whereas according to formulas (I, 37) and
(I, 47), Margoulis criterion is inversely proportional to
the Reynolds criterion to the power 0.2. The dependence

Knorre[22][*], the following relation was obtained for an air
stream flowing through a layer of regular spherical
particles

$$Nu = A \cdot Re^{m} \qquad (I, 49)$$

with m = 0.6. Here, the velocity involved in Reynolds'
criterion is calculated for the whole cross-section of
the layer (not for the free cross-section). For the
dimension d involved in the criteria of Nusselt and
Reynolds, one takes the diameter of a sphere.

The magnitude A in formula (I, 49) depends on
the separation of the layer. By separation, one means
the ratio of the volume of all intergranular spaces and
the total volume of the mass (without allowing for the
inner particle porosity). The dependence of the coefficient A in formula
(I, 49) on the separation
of the layer, obtained by
Bernshtein, is represented
in Figure 8. This graph
shows that, as a function
of the separation, the coefficient A passes through
a maximum. The reason is
that with increasing sep-
aration, the free surface
of the particles increases,
but, for a given value of
the velocity, the true
velocity calculated for the
whole cross-section of the
layer decreases.

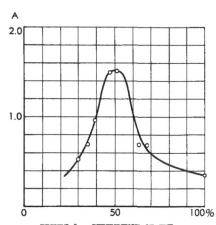

FIGURE 8.   DEPENDENCE OF THE
COEFFICIENT   A   OF BERNSHTEIN'S
FORMULA (I, 49) ON THE POROSITY
OF THE BED
Ordinates: coefficient A
Abscissas: porosity of the bed
in percent.

---

* See also: Colburn, A., Industrial Engineering Chemistry,
23 910 (1931).

Differential Equations of the Heat
Conductance and the Diffusion

The foregoing elementary exposition of the main
results of the theory of heat transfer in immobile and in
moving media are sufficient for the solution of many problems of macroscopic kinetics. However, in many cases we
shall have to use differential equations describing the
transfer of heat or of matter. These equations will be
reviewed in the following.

The equation of diffusion in an immobile medium
is of the form

$$\frac{\partial C}{\partial t} = \text{div D grad } C + q' \qquad\qquad (I, 50)$$

where C is the concentration, D the diffusion coefficient, and q' the density of the sources of substance,
i.e., the quantity of substances formed as a result of
chemical reactions in unit volume per unit time. The
symbols div and grad refer to vector-analysis opera-
tions termed, respectively, divergence and gradient.

If the diffusion coefficient D is constant and
q' = 0, equation (I, 50) becomes

$$\frac{\partial C}{\partial t} = D\Delta C \qquad\qquad (I, 50a)$$

where Δ = div grad is the Laplace operator which, in a
rectangular coordinate system is equal to the sum of the
second derivatives with respect to all three coordinates.

The equation of heat conductance in an immobile
layer is of the form'

$$c\rho\, \frac{\partial T}{\partial t} = \text{div } \lambda \text{ grad } T + q' \qquad\qquad (I, 51)$$

where T is the temperature, c the heat capacity, ρ
the density, λ the heat conductivity, and q' the

the density of the sources of heat, i.e., the quantity of heat evolved as a result of chemical reactions in unit volume per unit time.

If the heat conductivity λ can be considered constant, equation (I, 51) becomes

$$\frac{\partial T}{\partial t} = a\Delta T + \frac{q'}{c\rho} \qquad (I, 51a)$$

For stationary processes, the terms $\frac{\partial C}{\partial t}$ and $\frac{\partial T}{\partial t}$ are equal to zero.

In the presence of convection, the equations (I, 50) and (I, 51) must be supplemented by the terms v grad C and v grad T (where v is the flow velocity) and be solved simultaneously with the hydrodynamic equations.

Analysis of the Differential Equations by the Method of the Theory of Similitude

The theory of similitude enables one, without resorting to integration, to find the general forms of the solution save for the unknown function, i.e., to establish what dimensionless parameters should figure in the solution sought. To this end, it is necessary to express the equations in dimensionless variables by introducing as yardsticks for all the variables, magnitudes figuring in the conditions of the problem. In this way, all constant coefficients become magnitudes of the same dimension. By dividing the whole equation by one of these coefficients, we obtain a dimensionless equation in which dimensionless parameters play the role of constant coefficients. Its solution can contain, besides dimensionless variables, only such dimensionless parameters; all laws describing the given physical process can be represented by relations between these dimensionless parameters.

We will apply this method to the above differential equations. We introduce a natural length yardstick

d for which we take a dimension characteristic of the
syotem, and we replace the usual coordinates x by the
dimensionless coordinates

$$\xi = \frac{x}{d}$$

For the temperature and the concentration, we
introduce as natural yardsticks the characteristic temperature difference ΔT and the characteristic concentration
difference ΔC; the dimensionless temperature will then
be defined as

$$\theta = \frac{T - T_o}{\Delta T}$$

and the dimensionless concentration as

$$\zeta = \frac{C - C_o}{\Delta C}$$

where T_o and C_o are magnitudes arbitrarily taken as
the zero of temperature and of concentration, and figuring
in the conditions of the problem.

Finally, as a natural yardstick of time, we introduce the characteristic time τ and we define the
dimensionless time as

$$t' = \frac{t}{\tau}$$

After this transformation, the equations of heat conductance and of diffusion in an immobile medium (I, 50a) and
(I, 51a) take the form

$$\frac{\partial \theta}{\partial t'} = \frac{a\tau}{d^2} \cdot \frac{\partial^2 \theta}{\partial \xi^2} \; ; \quad \frac{\partial \zeta}{\partial t'} = \frac{D\tau}{d^2} \cdot \frac{\partial^2 \zeta}{\partial \xi^2}$$

the dimensionless concentration

$$\zeta = \frac{C - C_o}{\Delta C}$$

and the dimensionless coordinate

$$\xi = \frac{x}{d}$$

we transform (I, 10) into

$$\frac{q}{\Delta T} \frac{d}{\lambda} = - \left(\frac{d\theta}{d\xi} \right)$$

and (I, 11) into

$$\frac{q}{\Delta C} \frac{d}{D} = - \left(\frac{d\zeta}{d\xi} \right)$$

The left-hand members of these expressions are the familiar Nusselt criterion.

If the laws of Fourier and Fick are supplemented by convective members, their conversion to dimensionless variables will bring forth the Péclet criterion, and, in the same way, Reynolds' criterion will appear in hydrodynamic equations. Combination of these criteria will yield all the remaining similitude criteria discussed above.

Literature

1. ZELDOVICH. Teoriya goreniya i detonatsii gazov (Theory of combustion and detonation of gases) Moscow (1944).

2. FRANK-KAMENETSKII, Gorenie uglya (Combustion of coal). Uspekhi Khim. 7, 1277 (1938).

3. BUBEN and FRANK-KAMENETSKII, Zhur. Fiz. Khim. 20, 225 (1946).

4. DAMKOHLER, in Eucken-Jakob's Chemie Ingenieur, III, T. 1, 448 (1937).

5. FRANK-KAMENETSKII, KRENTSEL and ZBEREV, Khim.
    Promyshlennost' No. 1, 31 (1946).

6. FRANK-KAMENETSKII, Zhur. Fiz. Khim. 13, 738 (1939).

7. ZELDOVICH and SEMENOV, ZHUR. eksptl. teoret. Fiz.
    10, 1116 (1940).

8. KLIBANOVA and FRANK-KAMENETSKII, Acta Physicochim.
    18, 387 (1943).

9. LEVICH, Zhur. Fiz. Khim. 18, 335 (1944).

10. SEMENOV, Tsepnye reaktsii (Chain reactions). (1934).

11. FRANK-KAMENETSKII, Zhur. Fiz. Khim. 14, 695 (1940).

12. FRANK-KAMENETSKII, Kinetika slozhnykh reaktsii
    (Kinetics of complex reactions) Uspekhi Khim.
    10, 373 (1941).

13. FRANK-KAMENETSKII, Uspekhi Khim. 10, 544 (1941).

14. ZELDOVICH, Acta Physichochim. 1, 449 (1934).

15. GUKHMAN, Fizicheskie osnovy teploperedachi
    (Physical foundation of heat transfer) (1934).

16. KIRPICHEV, MIKHEEV, and EIGENSON, Teploperedacha
    (Heat Transfer) (1940).

17. MARGOULIS, W., Chaleur et Industrie, 134, 135, 269,
    352, (1931).

18. LÉVEQUE, Ann. des Mines, (12), 13, 201, 305, 381,
    (1928).

19. RUBINSHTEIN, Metod analogii s diffuziei; primenenie
    ego dlya issledovaniya teploperedachi v nachalnom
    uchastke truby (Method of analogy with diffusion
    and its application to the investigation of heat
    transfer in the initial portion of a tube).
    Article in the volume "Issledovanie protsessov
    regulirovaniya, teploperedachi i obratnogo
    okhlazhdeniya" (Investigation of processes of
    regulation, heat transfer and reverse cooling)
    (1938).

20. POHLHAUSEN, Z. Angew. Math. Mech. 1, 115 (1921).

21. VYRUBOV, Zhur. teckh. Fiz. 9, 1923 (1939).

22. "Issledovanie protsessov goreniya naturalnogo
    topliva" (Investigation of processes of
    combustion of natural fuel) edited by Knorre
    (in press).

CHAPTER II: DIFFUSIONAL KINETICS

       In the real course of a heterogeneous reaction in nature or in industry, the observed rate of reaction is determined partly by its true chemical kinetics at the surface and partly by the rate of transport of reactants to that surface through molecular or convective (or turbulent) diffusion. The study of the course of chemical reactions under such conditions is the object of diffusional kinetics.

       In principle, the problem can be approached by three methods, all three of which are used. The first consists in an exact analytical solution of the differential equation of the diffusion under boundary conditions which are determined by the kinetics of the reaction at the surface. In mathematical language these will be combined boundary conditions of the type

$$-D \text{ grad } C = kC^n$$

where the kinetics of the reaction is expressed by a formula of the type (I, 1).

       This first approach is used only in certain specific problems in the series of papers of Paneth and Herzfeld[1], Damköhler[2], Predvoditelev and Tsukhanova[3], Levich[4], and

Semenov[31]. These authors confined themselves to first-order
reactions. Each individual case, characterized by definite
geometric and physical conditions, becomes an independent
complex problem of its own. There can be no question of any
general results applicable to any geometric configuration
or any kind of motion of the gas.

The second method involves complete sacrifice of
an analytical solution and application of the theory of similitude not only to transfer processes but to the chemical
process itself. This method was sketched in a series of pa-
pers by Vulis[9] and was recently developed systematically by
D'yakonov[36].

In our own work extensive use was made of a third
method[5] which we call the "quasi-stationary method" or the
"method of uniformly accessible surface." This method of
approximation not only simplifies the calculation considerably but, which is even more important, permits isolation of
the physically essential limiting cases.

METHOD OF THE UNIFORMLY ACCESSIBLE SURFACE

This approximate but generally applicable method
of solution of diffusional kinetics consists in the following. It is assumed that the conditions of diffusional trans-
ference can be considered approximately independent of the
course of the reaction at the surface. This method is applicable to cases where all portions of the surface can be
considered equally accessible with respect to diffusion.
Such surfaces will be referred to as uniformly accessible.

The diffusion process can be described either by
integration of the diffusion equation with a simple ($C = 0$
at the surface) rather than a combined boundary condition or
by means of the experimental data of diffusion or heat exchange. This permits broad utilization of the methods and
the results of the theory of similitude.

Let the concentration of the reactant in space be

C and its concentration at the surface, where the reaction
takes place, be designated by C'. In the external problem, C will mean the concentration at infinite distance
from the surface and in the internal problem, the mean
volume concentration.

The rate of reaction at the surface depends on C'
and is given by the true chemical reaction kinetics at the
surface. Let the rate be expressed as a function of the
concentration of the reactant at the surface $f(C')$. In
the stationary or quasi-stationary state, the rate of reaction should be equal to the amount of reactant supplied
to the surface by molecular or turbulent diffusion, which
according to formula (I, 17) can be expressed

$$q = \beta(C - C') \qquad\qquad (II, 1)$$

where β is the diffusion velocity constant.

Equating the amount of substance consumed by reaction at the surface and the amount supplied by diffusion,
we get the algebraic equation

$$\beta(C - C') = f(C') \qquad\qquad (II, 2)$$

This reaction can be solved for C' for any concrete function $f(C')$. Substitution of the thus obtained
C' in (II, 1) or in $f(C')$ gives the desired value of
the overall reaction rate.

The diffusion velocity constant β can be found
either analytically or from the experimental data of
diffusion or heat transfer with the aid of the theory of
similitude according to which

$$\beta = \frac{Nu\ D}{d} = MV \qquad\qquad (II, 3)$$

First-Order Reaction

Consider the simplest case[5,6,7] where the reaction at the surface is of the first order

$$f(C') = kC' \qquad\qquad (II, 2a)$$

Equation (II, 2) then becomes

$$\beta(C - C') = kC'$$

and its solution is

$$C' = \frac{\beta C}{k + \beta} \qquad\qquad (II, 4)$$

The quasi-stationary reaction rate is

$$q = \frac{k\,\beta}{k + \beta}\,C \qquad\qquad (II, 5)$$

In this instance the macroscopic rate of reaction also follows the first order in reactant concentration. It can be written

$$q = k^*C \qquad\qquad (II, 6)$$

where

$$k^* = \frac{k\beta}{k + \beta} \qquad\qquad (II, 7)$$

This relation can be written in the instructive form

$$\frac{1}{k^*} = \frac{1}{k} + \frac{1}{\beta} \qquad\qquad (II, 7a)$$

which means that in this simplest case there is additivity of the reciprocals of the reaction rate and of the diffusion constants, i.e., of the diffusional and kinetic "resistances"[6,7].

II. DIFFUSIONAL KINETICS

Diffusional and Kinetic Region

Formula (II, 7) takes on a simple form in the two limiting ranges where one of the two magnitudes k and β is much greater than the other. If $k \gg \beta$, one has $k* \approx \beta$ and, according to (II, 4),

$$C' = \frac{\beta}{k} C \ll C$$

in which case the rate of the overall process is entirely determined by the rate of diffusion. if $k \ll \beta$, equation (II, 7) becomes $k* \rightharpoonup k$ and (II, 4) becomes $C' \approx C$, i.e., the overall rate is entirely determined by the true chemical kinetics of the reaction at the surface and does not depend on the conditions of the diffusion.

The same limiting ranges will exist also in cases of more complex types of the true kinetics at the surface.

If the diffusion resistance is much smaller than the chemical resistance, i.e., $\beta C \gg f(C)$ the macroscopic rate of reaction will coincide with the true reaction rate at the surface, i.e., will be $f(C)$, and the concentration of the reactant at the surface will be identical with its concentration in space, $C' \approx C$.

The degree of complexity of the dependence of the observed reaction rate on the reactant concentration will in that case be the same as for the true rate of the reaction, and the temperature dependence of the rate will be described by Arrhenius' law. The rate of reaction is entirely independent of the velocity of the gas stream and should be strictly proportional to the free reacting surface area.

This limiting range in which the true kinetics of the reaction can be measured directly has been termed[5] "the kinetic range".

In the other limiting case the diffusion resistance is much greater than the chemical resistance, $\beta C \ll f(C)$.

The observed rate of the overall process will under these
conditions be entirely determined by the rate of diffusion.
The concentration of the reactant at the surface will be
much lower than in the space, $C' \ll C$; the observed rate
of reaction can be expressed by

$$\beta C = \frac{Nu\ D}{d}\ C$$

This limiting range has been called the diffusional range[5].
In the diffusional range, the observed macroscopic rate of
reaction has nothing in common with the true kinetics at
the surface. All reactions in the diffusional range are of
the first order with respect to reactant concentration under
constant total pressure, and the order of the reaction with
respect to total pressure is different from its order with
respect to concentration.

In a medium at rest, Nu is a constant magnitude
depending only on the geometric configuration of the system.
At constant composition the concentration C will vary proportionally to the total pressure, and the diffusion coefficient D will vary inversely proportionally; as a result
the macroscopic rate of reaction will be independent of the
total pressure. In a gas stream, Nu will be a function
of the criterion

$$Re = \frac{Vd}{\upsilon} = \frac{Wd}{\mu}$$

where $V =$ linear velocity of the gas, $W =$ the mass velocity, $\upsilon =$ the kinematic viscosity, and $\mu =$ the dynamic
viscosity. In an ideal gas the dynamic viscosity is independent of the pressure. Consequently, if the total pressure
is varied at constant mass velocity W, the Reynolds number
Re will remain constant and hence also Nu and the observed
rate of reaction. As a result, in the diffusion range all
reactions must be of the first order with respect to reactant concentration under constant total pressure and must be

zero-order with respect to total pressure at constant composition and mass velocity.

    In the diffusional range, the rate of reaction de-
pends only little on the temperature as does the diffusion
coefficient; it obligatorily depends on the velocity of the
gas flow as does the Nusselt criterion namely, proportionally
to a 0.4 - 0.5 power of the velocity of the gas flow in
the external problem and an 0.8 power in the internal problem. In the diffusional range the rate of reaction does not
in any way depend on the specific chemical mechanism; rates
of different reactions in the diffusional range will differ
only insofar as the diffusion coefficients of the reactants
are different (in the case of reversible reactions, also
depending on the equilibrium conditions). The diffusional
range will be encountered under conditions favoring a high
rate of the chemical reaction and a low rate of diffusion:
at high temperatures, under high pressures, and at low velocities of the gas flow. In contrast, at low temperatures
under low pressures, and at high velocities of the gas flow
the kinetic range will be observed. These conditions are
necessary for the study of the true kinetics of a heterogeneous reaction.

    In between the diffusional and the kinetic range
lies an intermediate region where the course of the process
may be described by a solution of equation (II, 2). In particular, for a first-order reaction the rate in this intermediate region will be expressed by formula (II, 5).

Example of Chemical Reactions in the Diffusional Range

    In the diffusional range the concentration of the
reactant at the surface is much smaller than in the space
$C' \ll C$. Therefore, it follows immediately from (II, 1)
that the rate of reaction in this range is expressed by

$$q = \beta C \qquad\qquad (II, 8)$$

Thus, in the diffusional range, the diffusion velocity constant plays the role of the reaction rate constant.

Measurements of the rate of reaction in the diffusional range can tell nothing about the true kinetics of the reaction and its chemical mechanism. On the other hand, such measurements can well provide a method of investigation of diffusion processes, particularly of convective diffusion.

There are several methods to ascertain whether the reaction takes place in the diffusional range. The most reliable way is to carry out the reaction under geometric and hydrodynamic conditions sufficiently well-defined for accurate calculation of the absolute velocity of the diffusion. The observed rate of the reaction should be equal to the calculated rate of the diffusion. If the observed rate of the reaction is found to be substantially smaller than the calculated rate of diffusion, it can be concluded that the process lies within the kinetic range. Under no circumstance can the observed rate of reaction be greater than the rate of diffusion.

Very often the chemical process is studied under insufficiently defined hydrodynamic and geometric conditions. If, for example, the system is stirred with an agitator or a gas stream is blown through a layer of porous material, the absolute velocity of diffusion cannot be calculated. In such cases, the decision as to whether the process lies in the diffusional or the kinetic range is made on the basis of the dependence of its rate on different parameters. If the rate of the reaction depends on the velocity of the flow, this is evidence that the process takes place in the diffusional range. A strong temperature dependence, obeying the Arrhenius law, is evidence in favor of the kinetic range (if the diffusion takes place in the gas phase).

Very instructive results are obtained when the kinetics of a reaction are investigated under the conditions of the internal problem, i.e., with the flow of the reacting gas or liquid passing through a tube the inner surface of

which is the seat of the reaction. If the process is within
the diffusional range, one should observe the pattern characteristic of the laminar and the turbulent regions which
are separated by the hydrodynamic critical condition. In
the laminar range the rate of reaction depends only very
little on the velocity of the flow, whereas in the turbulent
region it is, in a first approximation, proportional to it.

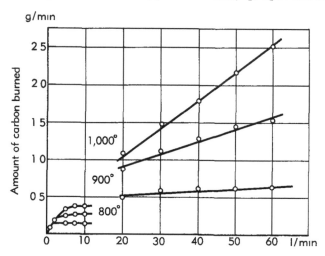

FIGURE 9.   RATE OF COMBUSTION OF A CARBON CHANNEL,
                AFTER TSUKHANOVA

Ordinates: amount of carbon burned (g/min)
Abscissas: velocity of the blast (l/min)
The curves refer to the temperatures of 800°,
900°, and 1000°C.

Correspondingly, in the laminar flow, the concentration of
the outgoing reaction products is approximately inversely
proportional to the velocity of the flow, whereas in the
turbulent it is practically independent thereof.

        The most thorough study of the course of a chemical
reaction in the diffusional range has been made for the processes of combustion of carbon and of solution of salts in
water and of metals in acids.

Combustion of Carbon

        It has been shown by many investigators[12, 13] that
the combustion of carbon above 1100 - 1300° C lies in the

diffusional range. Figures 9 and 10 represent the data obtained by Tsukhanova for the process of combustion of carbon under the conditions of the internal problem. The figures show clearly the regions of laminar and of turbulent combustion, separated by the hydrodynamic critical condition.

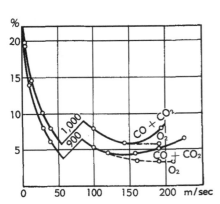

FIGURE 10.   COMPOSITION OF THE GAS IN THE COMBUSTION OF A CARBON CHENNEL, AFTER TSUKHANOVA

Ordinates: percent content of the
              combustion products in
              the gas.
Abscissas: velocity of the gas
              (m/sec).

The curves refer to the temperatures of 900° and 1000°C.

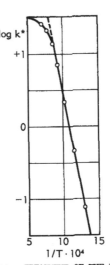

FIGURE 11.   TREATMENT OF THE EXPERIMENTAL DATA OF PREDVODITELEV AND TSUKHANOVA

Ordinates are log k* (effective rate constant).
Abscissas are 10^4 (1/T).
The curve is drawn according to our formula (II, 7). The circles represent the experimental data of Predvoditelev and Tsukhanova.

Predvoditelev and Tsukhanova[3] have investigated the range intermediate between the diffusional and kinetic regions (below 1100° C) and the kinetic region in laminar flow.

As has been shown by the author[14], the data of Predvoditelev and Tsukhanova can be very well described by formula (II, 5). In Figure 11, taken from[14], the effective rate constant k* is plotted as a function of the reciprocal temperature. The curve has been drawn with the aid of formula (II, 7); the circles correspond to the experimental data of Predvoditelev and Tsukhanova. The investigation of the

combustion of a carbon channel involves some experimental
difficulties. There is no doubt, however, that if good
experimental data are successfully obtained in a turbulent
flow, they will fit formula (II, 5), with the usual value
of the Nusselt criterion for a turbulent flow.

The reaction between carbon dioxide and carbon in
a carbon channel, in a laminar flow in the kinetic and the
intermediate ranges, has been studied by Vulis and Vitman[33].
These authors have evaluated their experimental data with
the aid of the analytical solution of Paneth and Herzfeld.

Vulis[34] has tabulated all available data on the
kinetics of the combustion of carbon which permit conclusions about the true chemical kinetics at the surface. The
results of this tabulation are given in Table 1. Vulis
applies Arrhenius' formula to the kinetic data. The table
gives the activation energies E and the preexponential
factor z for the reactions of carbon with oxygen and with
carbon dioxide.

The values of these kinetic magnitudes are different for different kinds of carbon. However, Vulis comes to
the conclusion that the ratio of the activation energies for
the two reactions is constant irrespective of the kind of
carbon; the activation energy for the reaction of carbon with
carbon dioxide is 2.2 times that for the reaction of carbon
with oxygen.

Vulis also finds that the logarithm of the pre-
exponential factor is a linear function of the activation
energy and that the rate constant can always be represented
by the empirical formula

$$k = k_o e^{\frac{E}{R}\left(\frac{1}{T^*} - \frac{1}{T}\right)}$$

where k_o has the same value for both reactions and is
equal to 31.6 cm/sec; $T^* = 1240^\circ$ K for the reaction of

carbon with oxygen, and $T* = 1840°$ K for the reaction of
carbon with carbon dioxide.

The only experimentally determinable characteristic
of a given kind of carbon is the activation energy; it is
sufficient to determine it for one of the two reactions.
All this treatment of the experimental data rests on the
assumption that the true chemical kinetics of the reaction
at the surface follows a first order. This assumption, however, lacks any theoretical or experimental substantiation.
And the porosity of the carbon is not taken into account.
Methods permitting an allowance for the effect of that factor
will be considered below.

The kinetics and the chemical mechanism of the reaction of carbon with carbon dioxide have been investigated
in detail by Semechkova and me[35] by measurements of the rate
of the reaction in the purely kinetic region with the result
that the order of reaction is actually lower than first.

Direct investigations of the kinetics of the reaction of carbon with oxygen are difficult because of large
thermal effects which result in considerable heating of the
surface. We have proposed in studying highly exothermic
reactions to use, in elucidating the nature of the reactions,
certain phenomena which depend upon the heat effect, such
as critical ignition conditions and extinction of reaction
at surfaces. Such matters will be discussed in detail in
Chapter IX.

In the work of Klibanova and the author[32], the investigation of the conditions of ignition of carbon filaments
in a stream of oxygen or air was used to establish the kinetic laws of the reaction between carbon and oxygen. The
main conclusion of that work is that the order of that reaction is not first, as is commonly assumed in the literature,
but fractional, lying between 1/3 and 1/2.

This result is borne out by data of direct measurements of the concentration of oxygen in the vicinity of the

KINETICS OF THE COMBUSTION OF CARBON (ACCORDING TO VULIS)

Reaction $C + CO_2$

Carbon	E kcal/mole	Z cm/sec	Authors
Electrode carbon	17.5	5×10^4	Vulis (channel)
Electrode carbon	20.3	1×10^3	Khitrin (particle)
Electrode carbon	22.0	2×10^5	Smith and Goodmunesen (particle)
Graphite	--	--	--
Electrode carbon	--	--	--
Electrode carbon	--	--	--
Graphite	--	--	--
Electrode carbon	25.5	1.1×10^3	Tsukhanova (channel)
Electrode carbon	26.5	2.5×10^6	Vulis (channel)
Electrode carbon	29.6	3.5×10^6	Nikolaev (particle)
Coke	31.5	1.6×10^7	Golovina (particle)
Coke	33.0	1.8×10^7	Nikolaev (particle)
Electrode carbon	35.5	2.2×10^7	Tiue et al.
Electrode carbon	35.5	5.6×10^7	Golovina (particle)
Electrode carbon	37.0	4×10^7	Blinov and Golovina (particle)

TABLE 1

Ratio $\dfrac{E_{C + CO_2}}{E_{C + O_2}}$	Carbon	E kcal/mole	Z cm/sec	Reaction $C + CO_2$ Authors
2.28	Electrode carbon	40.4	3×10^6	Vulis (channel)
--	Electrode carbon	--	--	--
--	Electrode carbon	--	--	--
--	Graphite	44.0	4×10^6	Meyers (plate)
--	Electrode carbon	44.25	6.9×10^6	Tsukhanova (channel)
--	Electrode carbon	51.2	7.9×10^7	Savvinov (particle)
--	Graphite	52.0	2.5×10^8	Meyers (plate)
2.23	Electrode carbon	52.0	1.6×10^9	Vulis (channel)
2.22	Electrode carbon	--	--	--
--	Electrode carbon	--	--	--
--	Coke	--	--	--
--	Coke	--	--	--
2.09	Electrode carbon	74.0	3.16×10^{10}	Dubinskii (particle)
--	Electrode carbon	--	--	--
--	Electrode carbon	--	--	--

burning carbon particle, if these data are analyzed from the point of view of diffusional kinetics of fractional-order reactions.

Fractional-Order Reactions

Let the rate of reaction be given as

$$f(C) = kC^m$$

where $m < 1$. Substituting this in formula (II, 2) we get

$$kC'^m = \beta(C - C') \qquad\qquad \text{(II, 9)}$$

where C is the concentration of the reactant in the volume, C' its concentration at the surface, and β · the diffusional velocity constant.

We convert equation (II, 9) to the dimensionless form

$$\frac{k}{\beta C^{1-m}} \left(\frac{C'}{C}\right)^m = 1 - \frac{C'}{C} \qquad\qquad \text{(II, 10)}$$

We designate the dimensionless magnitudes in formula (II, 10) by

$$\frac{C'}{C} = \zeta$$

$$\qquad\qquad \text{(II, 11)}$$

$$\frac{k}{\beta C^{1-m}} = \mu$$

Formula (II, 10) will then be written

$$\mu\zeta^m = 1 - \zeta \qquad\qquad \text{(II, 12)}$$

Solution of the algebraic equation (II, 12) will give the

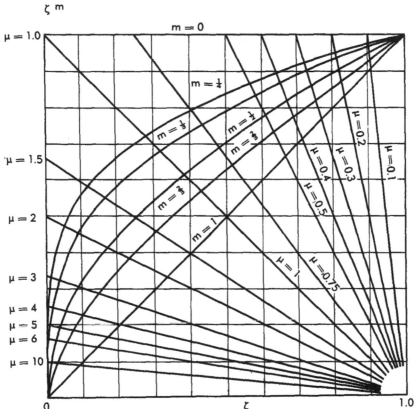

FIGURE 12. DIFFUSIONAL KINETICS OF A REACTION OF FRACTIONAL ORDER
The figure shows the graphic solution of equation (II, 12) $1 - \zeta = \mu\zeta^m$.
Abscissas represent the dimensionless concentration ζ; ordinates are ζ^m.
The parameters am are the different fractional orders of reaction. The
straight lines converging at the point $\zeta = 1$ represent the magnitude
$\frac{1 - \zeta}{\mu}$, for different values of μ. The abscissa of the point of intersection of the curve and the straight line gives the value of the
dimensionless concentration ζ as a function of μ for the given order
of reaction.

dimensionless concentration ζ as a function of the parameter

$$\zeta = f(\mu) \qquad\qquad (II, 13)$$

For the macroscopic rate of reaction we find

$$q = kC'^m = kC^m\zeta^m = kC^m[f(\mu)]^m \qquad (II, 14)$$

Figure 12 shows the graphic solution of equation (II, 12).
Abscissas are ζ, ordinates ζ^m. The straight lines passing
from the point $\zeta = 1$ are drawn by the formula

$$\frac{1}{\mu} (1 - \zeta)$$

their slopes correspond to different values of μ. The
curves starting at the origin of the coordinate system represent ζ^m as a function of ζ for different values of m.
The intersection of the curve and the straight line gives
the value of ζ satisfying equation (II, 12) as a function
of μ.

In the very interesting work of Parker and Hottel[11]
the concentration C' of oxygen at the surface of a burning
carbon particle was determined experimentally in microsamples
of gas taken directly at the surface, at different temperatures and velocities of the gas flow encompassed in the range
intermediate between the diffusional and the kinetic. We
compared[32] the experimental data of Parker and Hottel with
the values of C' calculated for different m by formula
(II, 12) and obtained additional data for the determination
of the order of the reaction.

The comparison is represented in Figure 13. The
abscissas are log μ, the ordinates are the corresponding
values of ζ calculated with the aid of the graph in Figure
12. The curves correspond to the different reaction orders
m. The curve for m = 1 corresponds to formula (II, 4).
The circles represent the experimental data of Parker and
Hottel, expressed in the dimensionless coordinates ζ and
μ. This transformation has been carried out in the following
way. The dimensionless concentration ζ was calculated directly as the ratio of the measured concentration at the
surface and the concentration in the volume. The value of
the parameter μ was calculated by formula (II, 11), putting

$$k = k_o e^{\frac{E}{RT_o^2} (T - T_o)}$$

$$\beta = \beta_0 \left(\frac{w}{w_0}\right)^a$$

where w is the velocity of the gas flow, and k_0 and β_0 the values of these magnitudes at an arbitrarily chosen temperature T_0 and velocity w_0. Designating the value of μ under these conditions by μ_0, we obtain from (II, 11)

$$\log\left(\frac{\mu}{\mu_0}\right) = \frac{E}{4.57 \ T_0^2} \ (T - T_0) - a \log \frac{w}{w_0} \quad (II, 15)$$

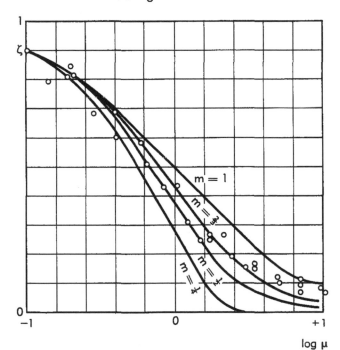

FIGURE 13.   TREATMENT OF THE EXPERIMENTAL DATA OF PARKER
AND HOTTEL

Ordinate represents the dimensionless oxygen concentration ζ, i.e., the ratio of the concentration at the surface and in the volume. Abscissas are $\log \mu$. The curves refer to different values of the order of reaction m. The circles correspond to the experimental data of Parker and Hottel.

Under the conditions of the experiments of Parker and Hottel a = 0.37. The value of E can be estimated from the dependence between w and T at constant ζ (i.e., at constant μ)

$$E = \frac{4.57 \, T_o^2}{T - T_o} \log \frac{w}{w_o} \qquad \text{(II, 16)}$$

Substitution of the experimental data of Parker and Hottel in formula (II, 16) gives the value $E = 41.1$ kcal/mole; the authors, by way of an entirely different method of treatment, have arrived at the value $E = 44$ kcal/mole.

Let us choose $T_o = 1000^o$ K and $w = 21.0$ cm/sec; under these conditions, according to the experimental data of Parker and Hottel, $\zeta = 0.9$. Hence, by comparison with the theoretical curves (Figure 13) we conclude that

$$\log \mu_o \approx -1$$

With this value of μ_o, we calculate with the aid of formula (II, 15) the values of μ for all experimental points of Parker and Hottel; with the adopted value of the activation energy, a temperature change by 100° corresponds to a change of $\log \mu$ by 0.62.

Data of Parker and Hottel thus recalculated are represented by circles in Figure 13. The authors themselves, along with other investigators, believed the dependence of the true rate of the reaction on the oxygen concentration to be of the first order, and the dependence of the concentration at the surface on that in the volume to obey formula (II, 4). They did not, however, attempt to test to what extent their own experimental data are in agreement with that assumption. Actually, the figure shows that all the points lie definitely below the curve of $m = 1$ (except for the highest value of μ, where C' is very small and the accuracy of ζ is poor) and are contained between the curves corresponding to $m = 1/2$ and $m = 2/3$. This is the same order of reaction as that derived from our ignition experiments. Thus, the data of Parker and Hottel provide evidence that the true order of the reaction between carbon and oxygen is less than unity.

The theoretical curves (Figure 13) reveal one
more interesting feature of fractional-order reactions.
The lower the reaction order, the earlier and the sharper is
the transition of the reaction into the diffusional region.
Whereas, at m = 1 the intermediate region extends to
very high values of μ, at m < 1 one finds asymmetric
curves falling rapidly to $\zeta = 0$. This means that even
at comparatively small μ, the concentration at the
surface becomes practically zero, and the reaction becomes governed by diffusion alone.

It is known that in the reaction between carbon
and oxygen the purely diffusional region, free from any
perturbation due to true chemical kinetics at the surface,
is easily observed. This is due to the true order of the
reaction being lower than unity, and can be quoted as one
more proof in favor of this conclusion.

Kinetics of Dissolution

Since the time of the classic work of Nernst and
Brunner, it has been shown by many investigators that dissolution processes of metals and salts in water and in aqueous
solutions are often governed by the laws of diffusion. In
most cases, dissolution processes are reversible. In the
diffusional range, the concentration established at the surface corresponds to saturation. The rate of dissolution is
proportional to the difference between that concentration
and the concentration in the bulk of the solution, the proportionality coefficient being the diffusion velocity constant. In most cases the kinetics of such reactions were
studied under conditions of agitation of the liquid, which
are quite undefined in the hydrodynamic sense. The extent
of motion of the liquid produced by the rotation of the agitator depends strongly on its construction; therefore, experiments of different investigators can be entirely non-
comparable from the point of view of absolute velocities.

The existence of a dependence of the rate of dissolution
on the number of revolutions of the agitator is always le-
gitimately taken as evidence that the process takes place
in the diffusional range. At higher rates of rotation, this
dependence usually becomes considerably weaker; this does
not, however, indicate yet a transition into the kinetic
range as is assumed by some investigators. As we have shown[7]
this can happen also in the diffusional range and be due to
the agitator at higher rotational rates failing to carry the
liquid along with it ("slipping effect").

In order to be able to draw definite conclusions
about absolute rates of dissolution from experiments with
agitation, one must first determine the hydrodynamic characteristic of the particular agitator used under the definite
hydrodynamic and geometric conditions of the experiment.
To this end, it is necessary to measure, with the given agitator, the rate of a particular dissolution process which
is definitely known to proceed in the diffusional range.

Noyes, and later King[15] and his collaborators,
studied the kinetics of dissolution under better defined
hydrodynamic conditions, namely with the cylindrical dis-
solving body rotating in the liquid. In this instance, as
in stirring with an agitator, very strong artificial turbulence is produced. At high rotational speeds, King observed
a direct proportionality between the rate of dissolution
and the number of revolutions, which corresponds to a constant value of the Margoulis criterion, independent of the
Reynolds criterion, as it should be in the limiting range
of pure turbulence. However, the observed dependence of
the rate of dissolution on the diffusion coefficient and on
the viscosity of the solution shows that the Margoulis criterion is quite substantially dependent on the Prandtl criterion.

If the viscosity of the solution is varied through
variation of the temperature or through suitable additions
(sucrose, colloids), the diffusion coefficient varies

inversely proportionally to the viscosity. Consequently,
Prandtl's criterion,

$$Pr = \frac{v}{D}$$

varies proportionally to the square of the viscosity. Thus,
a study of the diffusional kinetics of dissolution in high-
viscosity liquids permits the study of convective diffusion
in the limiting case of very high values of the Prandtl criterion as was first remarked by Levich[4]. This problem will
be considered in detail in Chapter V.

Of great interest is the study of processes of
dissolution under hydrodynamically better defined conditions
of the internal problem. The first studies of this type
were carried out by Uchida and Nakayama[16] who followed the
dissolution of a copper tube in cuproammoniacal solutions.
The extraordinary complexity of this process renders treatment of their data very difficult. The process consists
essentially of two superposed reactions, one of which is
limited by the diffusion of oxygen, the other, by the diffusion of the divalent Cu^{++} ion. The difficulty of accurately separating these two reactions does not permit calculation, from the data of Uchida and Nakayama, of the correct
values of the absolute rate of the diffusion.

Aidarov, Buben, and the author[17] studied the dissolution of a copper tube in nitric acid under the conditions
of the internal problem. Through additions of nitrite, which
catalyzes the reaction, and of hydrogen peroxide, which sup-
presses gas evolution, it was possible to have the process
take place in the purely diffusional range. The results
are represented in Figure 14, where the values of the Nusselt
criterion calculated from the experimental data are plotted
as a function of the Péclet criterion. The vertical line
gives the value of Péclet corresponding to the point of
critical hydrodynamic conditions. Smaller values of Pe corre-

spond to the laminar, higher values of Pe, to the turbulent region. The points represent the experimental data. The full curve in the laminar region was drawn with the aid of Lévêque's formula.

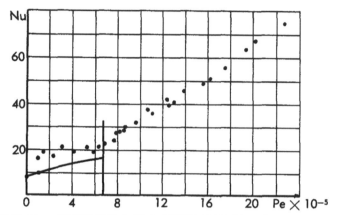

FIGURE 14.   DISSOLUTION OF A COPPER TUBE IN NITRIC ACID, ACCORDING TO THE EXPERIMENTS OF AIDAROV, BUBEN, AND FRANK-KAMENETSKII

Ordinates are the diffusional Nusselt criterion; abscissas are the diffusional Péclet criterion. The vertical line corresponding to the critical value of the Reynolds criterion. To the left of it lies the laminar, and to its right, the turbulent region. The full curve in the laminar region was drawn according to Lévêque's law (I, 39). The points represent the experimental results.

In the laminar region the rate of dissolution (Nusselt criterion) is only little dependent on the velocity of the flow (Péclet criterion).   This dependence is close to that which could be expected from Lévêque's formula, but the absolute values of the Nusselt criterion are markedly higher than those given by that formula.   In the turbulent region the rate of dissolution is directly proportional to the velocity of the flow, which corresponds to a constant value of the Margoulis criterion.

All experiments of diffusional-hydrodynamic kinetics of dissolution, made under a variety of hydrodynamic conditions, can be evaluated by way of calculation of the Margoulis criterion which represents the ratio of the diffusion velocity constant and the characteristic linear velocity. The latter, under the conditions of the internal problem, means the velocity of the flow, and in stirring with an agitator or rotation of the solid, the peripheral linear velocity of rotation.

Table 2 summarizes the experimental data of kinetics of dissolution obtained by different experimentors, and the values of the Margoulis criterion derived therefrom.

TABLE 2

Kinetics of Dissolution

No.	Object of Study	M	D cm^2/min	Pr	M.Pr
1	Cu in $FeCl_3$ (Buben, Kostyunin and Frank-Kamenetskii)	3.33×10^{-5}	0.42×10^{-3}	2000	6.66×10^{-2}
2	Cu in $FeCl_3$ + $CaCl_2$ (Buben, Kostyunin and Frank-Kamenetskii)	6.0×10^{-7}	0.08×10^{-3}	40000	2.4×10^{-2}
3	Fe in $FeCl_3$ (Abramson and King)	3.13×10^{-5}	0.215×10^{-3}	2500	8.3×10^{-2}
4	Fe in $HClO_4$ (Abramson and King)	11.1×10^{-5}	3.60×10^{-3}	149	1.65×10^{-2}
5	Zn in CH_3COOH (King and Schack)	5.5×10^{-5}	1.76×10^{-3}	706	3.9×10^{-2}
6	Gypsum in H_2O (Buben, Valova and Frank-Kamenetskii)	5.16×10^{-5}	0.78×10^{-3}	788	3.5×10^{-2}
7	Cu in HNO_3 (Aiderov, Buben and Frank-Kamenetskii)	3.3×10^{-5}	1.53×10^{-3}	400	1.3×10^{-2}
				mean	4.0×10^{-2}

Under the simplest assumptions of the existence of a laminar
sublayer of finite thickness at the solid surface, the
Margoulis criterion should, all other conditions being equal,
be inversely proportional to the Prandtl criterion. That
is why we have also included in the table the product of
the Margoulis and of the Prandtl criterion.

The problem of the hydrodynamic interpretation of
diffusional-kinetics data of dissolution will be considered in more detail in Chapter V.

Heterogeneous Termination in Chain Reactions

In kinetics of chain reactions, a major role is
played by processes of heterogeneous termination of reaction chains when the active species is annihilated through
a heterogeneous reaction taking place at the solid surface.

Most often such a termination occurs at the wall
of the reaction vessel. In a recent paper of Semenov[31],
instances are also considered where the termination takes
place at the surface of a metallic rod introduced into a
glass vessel, or at the surface of a spherical metal powder
particle suspended in the gas. The latter model is applied
to the interpretation of the action of metallic antideto-
nators on the oxidation of the fuel in internal-combustion
engines. An experimental confirmation of Semenov's conclusions was given in the work of Nalbandyan and Shubina[37].

An unsatisfactory and possibly misleading terminology is often used in the literature on chain theory. In
cases where the rate of the termination reaction is determined by the diffusion of the active centers to the surface, it is said that "termination occurs at each collision
between the active center and the wall". Actually, the relation between the rate of the reaction and the number of
collisions is entirely irrelevant here. What is essential
is that the termination takes place in the diffusional range.
The contrary limiting situation is usually described in the

literature by the statement "the probability of the termination is very small" where, again, one does not say
relative to what it is small. Actually it means that the
termination reaction takes place in the kinetic range.

Only in a recent paper of Semenov[31] does one find
a correct criterion which permits one to decide whether the
chain termination reaction takes place in the diffusional
or in the kinetic range. Semenov prefers to express this
criterion, not by the macroscopic characteristics of the
diffusion and the kinetics, but by gas-kinetic magnitudes.
In our notation, the process takes place in the kinetic
range when $k \ll \beta$

In the case of diffusion in a medium at rest, we
have

$$\beta \approx \frac{D}{d}$$

where d is the diameter of the vessel.

According to the kinetic gas theory,

$$D \approx \lambda U; \quad k \approx \epsilon U$$

where λ is the mean free path, U the root mean square
velocity of molecular motion, ϵ the probability of termination, i.e., the ratio of the rate of termination and the
number of collisions.

Bearing in mind that the mean free path is inversely
proportional to the pressure, Semenov writes, moreover

$$\lambda = \frac{\lambda_0}{p}$$

and obtains the criterion of the termination taking place
in the kinetic range in the form

$$\epsilon p d < \lambda_0$$

The diffusional kinetics in this case has certain particular
features since the heterogeneous chain termination reaction

takes place at the wall of a closed vessel containing a motionless gas. The velocity of diffusion is determined by the
concentration distribution of the reacting substance (active
product) in this gas which, in turn, depends on the boundary
conditions, i.e., on the true kinetics at the surface.

This case of diffusional kinetics can be of considerable interest not only in connection with the termination of reaction chains but also in view of other possible
applications. It deserves special consideration.

Reaction at the Wall of a Closed Vessel

Let us consider the diffusional kinetics of a
reaction taking place at the inner surface of a closed vessel containing a motionless gaseous mixture. In the absence
of convection, a complete analytical discussion of the process is possible by way of integration of the diffusion
equations in a motionless medium. Practically, this case
of diffusional kinetics is the basis of the theory of heterogeneous termination of reaction chains and was examined
from this point of view by a number of authors. The most
complete treatment was given in the work of Semenov[31].

We shall show that our method of discussion permits a substantial simplification and a more instructive
solution of the problem. We introduce the diffusion velocity
constant β , defined as the ratio of the diffusion flow
and the mean concentration difference. In a first approximation, we consider the value of the diffusion velocity constant independent of the rate of reaction and the same for
the stationary and the nonstationary problems. This assumption is not accurate since the concentration distribution
in the gas depends on the boundary conditions. We shall
test it by comparison with the accurate analytical calculation of Semenov.

In nonstationary diffusional processes the value
of the diffusional velocity constant changes with time.

At the initial moment when a very steep concentration gradient exists at the wall, diffusion is very intense; subsequently, a continuous concentration distribution is established and the diffusion process is slowed down.

After the lapse of a sufficiently long time from
the initial moment, the process goes over into the so-called
quasi-stationary stage, characterized by a constant value
of the velocity of diffusion.

Mathematically, the nonstationary diffusion process
in a motionless medium is described by the differential equation (I, 50a). The solution of this equation can be found
by the usual Fourier method as a sum of the partial integrals
representing the product of the function $T(t)$, depending
only on the time, and the function $X(x)$, depending only
on the coordinates (x designates the totality of all three
spatial coordinates).

C designates the value of the concentration at
the boundary, i.e., at the wall of the vessel and φ the
concentration difference

$$\varphi = C - C'$$

We seek φ as a sum of expressions of the form

$$\varphi_1 = T_1(t) \, X_1(x) \qquad\qquad (II, 17)$$

Substituting (II, 17) in (I, 50a), we get

$$\frac{1}{T_1} \cdot \frac{\partial T_1}{\partial t} = \frac{D}{X_1} \Delta X_1 \qquad\qquad (II, 18)$$

Inasmuch as the left-hand member of this equation depends
only on the time and the right-hand member only on the coordinates, each must be equal to a constant number which will
be designated by $-k_1^2$ the number k_1 is called the characteristic number of the equation.

We then get two ordinary differential equations for the functions T and X:

$$\frac{dT_1}{dt} = - k_1^2 T_1 \qquad\qquad (II, 19)$$

$$D\Delta X_1^2 = - k_1^2 X_1 \qquad\qquad (II, 20)$$

The solution of (II, 19) is of the form

$$T_1 = A_1 e^{-k_1^2 t}$$

Solutions of (II, 20) satisfying the boundary conditions are called fundamental functions. The requirement that the solutions satisfy the boundary conditions determines all the possible values of the characteristic number k_1. We can now write the general integral of equation (I, 50a) in the form

$$\varphi = \Sigma_1 A_1 e^{-k_1^2 t} X_1(x) \qquad\qquad (II, 21)$$

where X_1 are the fundamental functions, and k_1, the corresponding characteristic numbers.

After the lapse of a sufficiently long time from the initial moment, the exponential factors

$$e^{-k_1^2 t}$$

become very small and decrease very rapidly with increasing k_1. Consequently, after sufficiently long times, all terms in the series (II, 21), except the first, can be disregarded and one may put with sufficient approximation

$$\varphi \approx A_1 e^{-k_1^2 t} X_1(x) \qquad\qquad (II, 22)$$

where k_1 is the first characteristic number and X_1 the corresponding fundamental function.

The stage at which the approximate solution (II, 22) is applicable is termed the "quasistationary stage" of the process. At that stage the concentration distribution remains similar to itself, and the concentration difference at any given point decreases exponentially with time.

As can be seen from (II, 22), at the quasistationary stage of the process we can write for both the concentration difference at any given point and for the mean concentration difference

$$\varphi \sim e^{-k_1^2 t} \qquad\qquad \text{(II, 23)}$$

For this stage, one can introduce the characteristic diffusion time τ i.e., the time within which the concentration difference decreases by a factor of e. From (II, 23) one sees that

$$\tau = \frac{1}{k_1^2} \qquad\qquad \text{(II, 24)}$$

with this definition of τ, one gets

$$\varphi \sim e^{-\frac{1}{\tau}} \qquad\qquad \text{(II, 25)}$$

Differentiation of (II, 25) gives

$$\frac{\partial \varphi}{\partial t} = -\frac{\varphi}{\tau} \qquad\qquad \text{(II, 26)}$$

which holds for both the concentration difference at any point and for the mean concentration difference.

On the other hand, a change of the mean concentration is related with the amount of substance diffused, i.e., with the diffusion flow by the condition of balance of matter over the whole volume of the vessel.

II. DIFFUSIONAL KINETICS

$$\omega \frac{\partial \overline{\varphi}}{\partial t} = - qS \qquad \text{(II, 27)}$$

where ω is the volume of the vessel, S its wall surface area, q the diffusion flow; the upper bar indicates mean value over the whole volume of the vessel.

We introduce the diffusion velocity constant defined as the ratio of the diffusion flow and the mean concentration difference at a given moment

$$\beta = \frac{q}{\overline{\varphi}} \qquad \text{(II, 28)}$$

Then, from (II, 26), (II, 27), and (II, 28) we get

$$\frac{\overline{\varphi}}{\tau} = q \frac{S}{\omega} \qquad \text{(II, 29)}$$

$$\beta = \frac{\omega}{S\tau} \qquad \text{(II, 30)}$$

By dimensional considerations one finds, as we have seen in Chapter I, that the characteristic time ought to be proportional to

$$\frac{d^2}{D}$$

where d is the dimension of the vessel and D the diffusion coefficient. The same conclusion is reached also by analysis of equation (II, 20). Since the Laplace operator is the sum of the second derivatives with respect to the coordinates, we can by introducing instead of the coordinates x the dimensionless coordinates

$$\xi = \frac{k_1 x}{\sqrt{D}}$$

put equation (II, 20) in a form not containing any parameters. Consequently, the functions X_1 should depend only on the magnitude ξ;   the boundary conditions take the form

$$\frac{k_1 d}{\sqrt{D}} = \mu_1 \qquad\qquad (II, 31)$$

where μ_1 are dimensionless characteristic numbers which depend only on the geometric shape of the vessel, and d is the diameter of the vessel.

Combining (II, 31) with (II, 24), we get

$$\tau = \frac{d^2}{\mu_1{}^2 D} \qquad\qquad (II, 32)$$

in agreement with the result found by dimensional considerations.

Substituting the expression (II, 32) in (II, 30), we find finally for the diffusion velocity constant,

$$\beta = \frac{D}{d^2} \mu_1{}^2 \frac{\omega}{S} \qquad\qquad (II, 33)$$

In the diffusional region, the effective reaction rate constant is directly equal to the thus determined diffusion velocity constant. In the kinetic region it is governed by the true kinetics at the surface. In the intermediate range we can make use of our approximate quasistationary method.

Bearing in mind that the value of the diffusion velocity constant does not change on passing from the diffusional into the intermediate region, one can find the concentration at the surface and the rate of the reaction from (II, 2). Specifically, for a first-order reaction, we get the already known result (II, 7)

$$k^* = \frac{k \beta}{k + \beta}$$

or

$$\frac{1}{k^*} = \frac{1}{k} + \frac{1}{\beta}$$

II. DIFFUSIONAL KINETICS

In connection with the theory of chain termination at the walls of the reaction vessel, analytical calculations of the course of a first-order reaction at the walls of a closed vessel were carried out by Kassel and Storch, and Lewis and von Elbe. The most detailed exposition of the problem is to be found in a recent article of Semenov[31].

For the diffusional range, integration of equation (II, 20) is a very elementary operation. The fundamental functions are, for a plane vessel, cosine; for a cylindrical vessel, Bessel functions; and for a spherical vessel,

$$\frac{\sin x}{x} .$$

Table 3 summarizes the main results of the calculation.

TABLE 3

Diffusional Kinetics at the Walls
of a Closed Vessel

	μ_1	τ	β	$\frac{\omega}{S}$
Plane vessel	π	$\dfrac{d^2}{\pi^2 D}$	$\dfrac{\pi^2}{2}\dfrac{D}{d}$	$\dfrac{d}{2}$
Cylindrical vessel	$2\mu_0$	$\dfrac{d^2}{4\mu_0^2 D}$	$\mu_0^2\dfrac{D}{d}$	$\dfrac{d}{4}$
Spherical vessel	2π	$\dfrac{d^2}{4\pi^2 D}$	$\dfrac{2}{3}\pi^2\dfrac{D}{d}$	$\dfrac{d}{6}$

Here, $\mu_0 = 2.4048$ is the first root of the Bessel function of zero order.

For a cylinder of finite length L and diameter d, the quasistationary diffusion time is found from

$$\frac{1}{\tau} = \left(\frac{\pi^2}{L^2} + \frac{4\mu_o^2}{d^2} \right) D$$

In the intermediate region, one can in a first approximation make use of our method of uniformly accessible surface and put

$$k^* = \frac{k\beta}{k + \beta}$$

with the values of β taken from the above table.

An accurate solution of the problem is given by Semenov[31]. He introduces the auxiliary magnitudes μ and u_1, which in our notation are expressed by

$$\mu = \frac{kd}{2D} \qquad\qquad (II, 34)$$

$$u_1^2 = \frac{1}{4} \frac{d^2}{D} \frac{S}{\omega} k^* \qquad\qquad (II, 35)$$

Accurate integration of the stationary diffusion equation shows that the magnitude u_1 is the first positive root of the transcendental equation which is

for a plane vessel

$$\cotg u = \frac{u}{\mu} \qquad\qquad (II, 36)$$

for a cylindrical vessel

$$\frac{I_o(u)}{I_1(u)} = \frac{u}{\mu} \qquad\qquad (II, 37)$$

(where I_o and I_1 are Bessel functions of zero and first order)

for a spherical vessel

$$1 - u \cot u = \mu \qquad (\text{II, 38})$$

Solutions of these equations for cylindrical and spherical vessels are given in Semenov's article in the form of graphs which show the dependence of u_1^2 on μ.

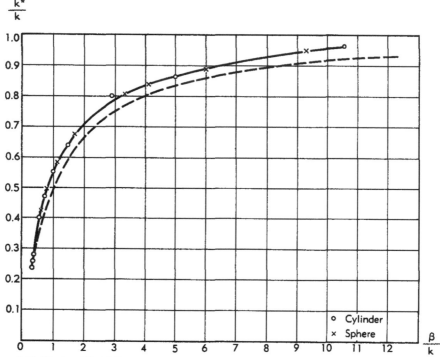

FIGURE 15. DIFFUSIONAL KINETICS AT THE WALL OF A CLOSED VESSEL
Ordinates are the ratio $\frac{k^*}{k}$ of the effective rate constant k^* and the true rate constant of the chemical reaction k. Abscissas are the ratios $\frac{\beta}{k}$ of the diffusion velocity constant β and the rate constant of the chemical reaction k. The crosses represent the results of the calculations of Semenov for a cylinder, and the circles, for a sphere. The full line was drawn through the points and expresses the result of the accurate calculation. The dotted line was drawn according to formula (II, 7), i.e., corresponds to the method of uniformly accessible surface.

Recalculation of these graphs for the dependence of $\frac{k^*}{k}$ on $\frac{\beta}{k}$ gives the results represented by the solid line in Figure 15. Points corresponding to the cylindrical

and the spherical vessel lie exactly on the same curve. The
dotted line in this figure was drawn by formula (II, 7),
which corresponds to our method of uniformly accessible surface. As the figure shows, it comes quite close to the
accurate solution.

Diffusional Kinetics of Complex Reactions

Up to this point we have considered the simplest
case of reactions, the rate of which depends on the concentration of one reactant only. We shall now discuss cases
of more complex reactions, limiting ourselves to the diffusional range only.

Case of Several Diffusing Substances

Let the reaction involve several diffusing particles linked by one stoichiometric equation

$$v_1A_1 + v_2A_2 + \dots$$

where A_1 , A_2 , $\dots$ are the chemical symbols of the original
reactants and where v_1 , v_2 , $\dots$ are the stoichiometric coefficients. In the quasistationary course of the process,
all initially present substances should undergo diffusion
to the surface in equivalent amounts, i.e., the condition
of stoichiometry of the diffusion flows should be fulfilled,

$$\frac{q_1}{v_1} = \frac{q_2}{v_2} = \dots = \frac{q_i}{v_i} = \dots \qquad (II, 39)$$

Evidently, no matter how great the rate of the reaction at
the surface, the concentration of all reactants cannot, if
condition (II, 39) is to be fulfilled, become small at the
same time. Let us assume that for all substances, their
concentrations at the surface C'_i have become small compared with the concentrations in the volume C_i . The diffusion flows will then be expressed as

$$q_1 = \beta_1 C_1 \qquad\qquad (\text{II, }40)$$

and condition (II, 39) can be fulfilled only in one special case, namely at such a composition of the mixture that

$$\frac{\beta_1 C_1}{v_1} = \frac{\beta_2 C_2}{v_2} = \ldots = \frac{\beta_i C_i}{v_i} = \ldots \qquad (\text{II, }41)$$

If all diffusion velocity constants β_i are equal, it will be a mixture in which the concentrations of the substances are proportional to their stoichiometric coefficients, i.e., a stoichiometric mixture. If the diffusion velocity constants are different, the condition (II, 41) will be fulfilled at one definite composition of the mixture, different from the stoichiometric.

According to formula (II, 3),

$$\beta = \frac{\text{Nu } D}{d}$$

In the case of diffusion in a motionless medium, the values of both d and Nu will be the same for all substances (in a motionless medium the Nusselt criterion depends only on the geometric shape of the system). Consequently, in diffusion in a motionless medium the β_i will be proportional to the D_i and conditions (II, 41) will be reduced to

$$\frac{D_1 C_1}{v_1} = \frac{D_2 C_2}{v_2} = \ldots = \frac{D_i C_i}{v_i} = \ldots \qquad (\text{II, }41\text{a})$$

In the presence of convection, the value of Nusselt's criterion depends on the Prandtl criterion and, consequently, also on the diffusion coefficient; therefore, in this case, the values of the Nusselt criterion will be different for different substances. Let us represent, as usual, the dependence of the Nusselt on the Reynolds and Prandtl criteria in the form (I, 41)

$$Nu = k \ Re^m Pr^n$$

We note that

$$Re = \frac{Vd}{\upsilon}$$

has the same value for all substances since all the magnitudes included in this parameter are characteristic of the system as a whole. On the other hand, the (diffusional) Prandtl criterion

$$Pr = \frac{\upsilon}{D}$$

is inversely proportional to the diffusion coefficient of the given substance, inasmuch as the kinematic viscosity υ is a characteristic property of the whole mixture and not of an individual component. Consequently, in the presence of convection, the values of the Nusselt criterion for different substances will be inversely proportional to their diffusion coefficients to the power n. Hence, condition (II, 41) gives

$$\frac{D_1^{1-n}C_1}{\upsilon_1} = \frac{D_2^{1-n}C_2}{\upsilon_2} = \ldots \frac{D_i^{1-n}C_i}{\upsilon_i} = \ldots \quad (II, 41b)$$

Condition (II, 41) will be referred to as the condition of diffusional stoichiometry; it amounts to the requirement that the velocities of diffusion of the reactants be proportional to their stoichiometric coefficients. Only in a mixture of a composition satisfying the condition of diffusional stoichiometry, can the concentrations of all reacting substances at the surface at the same time become small as compared with the concentrations in the volume. In an immobile medium and in the presence of convection, this will occur at different compositions of the mixture; in the first case the condition of diffusional stoichiometry is reduced to

(II, 41a) and in the second, to (II, 41b). Only in the particular case when the diffusion coefficients of all reactants
are equal, does the condition of diffusional stoichiometry
coincide with the usual stoichiometric composition of the
mixture.

If the condition of diffusional stoichiometry is
not fulfilled, the rate of the process will always be determined by the diffusion of one of the reacting substances.
Were the flows of the reactants not to satisfy the equivalence condition (II, 39) and were one substance to arrive
at the surface in an amount greater than that required by
the stoichiometry, this substance would unavoidably accumulate at the surface, no matter how great the rate of the
reaction. Its concentration C'_1 would grow until the diffusion flow

$$q_1 = \beta_1(C_1 - C'_1)$$

would begin to fulfill the condition of stoichiometry of the
flow. Consequently, for all substances except that for which
the magnitude

$$\frac{\beta_1 C_1}{v_1}$$

has the smallest value, the concentration C'_1 at the surface will not be small in comparison with the concentration
in the volume. The concentrations of these substances at
the surface can be determined from the condition of stoichiometry of the flow

$$\frac{\beta_1}{v_1} (C_1 - C'_1) = \left(\frac{\beta C}{v}\right)_{min} \qquad (II, 42)$$

where

$$\left(\frac{\beta C}{v}\right)_{min}$$

is the smallest of the

$$\frac{\beta_1 C_1}{v_1}$$

The diffusion velocities of all substances, except that for which $\frac{\beta C}{v}$ is minimum, do not depend on either the concentrations of these substances or on their diffusion coefficients. They are determined only by the stoichiometry. In other words, the whole course of the process is determined exclusively by the diffusion of the substance for which the magnitude $\frac{\beta C}{v}$ is minimum. According to the foregoing, in the case of diffusion in a motionless medium, β can be replaced by the diffusion coefficients D and in the presence of convection, by the magnitudes D^{1-n}.

These ideas are verified by experimental data of kinetics of dissolution of metals in acids in the presence of oxidants obtained by Pletenev[24] and by King[15] and his collaborators. The observed dependence of the velocity of dissolution on the concentration of the oxidant (depolarizer) is represented in Figure 16. At low concentrations of the oxidant, the rate of dissolution is proportional to its concentration and does not depend on the concentration of the acid.

This holds up to a certain concentration of the oxidant, corresponding to the diffusional stoichiometry. When this is attained, the rate of dissolution ceases to depend on the concentration of the depolarizer and becomes proportional to the concentration of the acid. In the first instance, the process is limited by the diffusion of the depolarizer, in the second, by the diffusion of the acid. A similar situation was observed in the hydrogenation of unsaturated compounds in the liquid phase: the rate of the hydrogenation was found independent of the concentration of the substance undergoing hydrogenation and was found proportional to the pressure of hydrogen.

Thus, in all cases where several substances linked by one stoichiometric equation take part in the reaction, the rate is always limited by the diffusion of one of these substances, that for which the magnitude $\frac{\beta C}{v}$ has the smallest value.

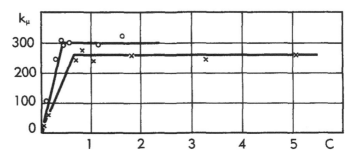

FIGURE 16.   DEPENDENCE OF THE RATE OF DISSOLUTION ON THE CONCENTRATION OF THE OXIDANT, AFTER PLETENEV
Ordinates are the rates of dissolution. Abscissas are the concentration of the oxidant.

This, of course, does not apply to the case where the surface is the seat of several independent parallel reactions, the rates of which can be determined by the diffusion of different substances.

In the instance of the dissolution of copper in cuproammoniacal solutions, studied by Yamazaki and Uchida and Nakayama[16], there are apparently two superposed parallel processes: dissolution of copper under the action of oxygen

$$Cu + \frac{1}{2} O_2 + H_2O = Cu^{++} + 2OH^-$$

the rate of which is determined by the diffusion of oxygen, and dissolution by divalent copper ions,

$$Cu + Cu^{++} = 2Cu^+$$

followed by oxidation of the Cu^+ ion in the volume. If the latter process can take place in the diffusional range, its rate will be limited by the diffusion of Cu^{++} ions. In this case, the diffusion of oxygen and of Cu^{++} are both essential for the rate of dissolution of the copper. This would be entirely impossible if these substances were tied together by one single stoichiometric equation.

Reversible Reactions

The rate of a reversible reaction in the diffusional range does not depend on its true kinetics but on the thermodynamic equilibrium. In many cases, a close tie was observed experimentally between the rate of reaction and the thermodynamic equilibrium. It is quite probable that often this is due to the reaction taking place in the diffusional region.

As an example let us consider the simplest case of a reversible reaction in which one molecule of the original reactant gives one molecule of the product $A \rightleftarrows B$. Let us designate the concentration of A and B in the volume by C_A and C_B and designate their concentrations at the surface by C_A' and C_B'. If both the forward and the reverse reaction take place in the diffusional region, the equilibrium set up at the surface will be determined by the mass-action law

$$\frac{C'_A}{C'_B} = K \qquad\qquad (II, 43)$$

where K is the equilibrium constant. On the other hand, in the quasistationary course of the process, the rate of diffusion of the initial reactant to the surface must be equal to the rate of the reverse diffusion of the reaction product away from the surface.

With β_1 designating the diffusion velocity constant of the initial reactant A and B_2 designating that of the product B, we get

$$\beta_1(C_A - C_A') = \beta_2(C_B' - C_B) \qquad\qquad (II, 44)$$

Conditions (II, 43) and (II, 44) give us two equations for the determination of the two unknown magnitudes C_A' and C_B'. Solving them, we get

$$C'_A = \frac{\beta_1 C_A + \beta_2 C_B}{\beta_1 + \beta_2 K}; \qquad C'_B = \frac{\beta_1 C_A + \beta_2 C_B}{\frac{\beta_1}{K} + \beta_2}$$

The rate of reaction, equal to the rate of diffusion, is

$$q = \beta_1 (C_A - C'_A) = \beta_2 (C'_B - C_B) = \frac{\beta_1}{K + \frac{\beta_1}{\beta_2}} (KC_A - C_B)$$

$$(II, 45)$$

This expression can be put in a very elegant form with the aid of the so-called "maximum work" of the reaction

$$A = RT \ln K - RT \ln \frac{C_B}{C_A} = RT \ln \frac{KC_A}{C_B} \ldots \quad (II, 46)$$

where R is the gas constant, and T the absolute temperature.

The magnitude A is equal to the decrease of the free energy of the system as a result of the reaction

$$A = -\Delta F$$

In order for the reaction to proceed, the magnitude A must be positive.

With the aid of the definition of the magnitude A, we can put formula (II, 45) into the form

$$q = \frac{\beta_1 C_A \left(1 - e^{-\frac{A}{RT}}\right)}{1 + \frac{\beta_1 C_A}{\beta_2 C_B} e^{-\frac{A}{RT}}} \qquad (II, 47)$$

We shall consider the limiting cases when the system is far from or close to equilibrium:

Far from equilibrium

$$A \gg RT; \qquad e^{-\frac{A}{RT}} \ll 1$$

and formula (II, 47) tends to

$$q \cong \beta_1 C_A \qquad\qquad (II, 48)$$

i.e., the rate of the process is determined by the diffusion of the initial substance, as it should be in the diffusional region.

Near equilibrium

$$A \ll RT$$

we can develop

$$e^{-\frac{A}{RT}}$$

into a series

$$e^{-\frac{A}{RT}} \approx 1 - \frac{A}{RT}$$

and (II, 47) will give

$$q \cong \frac{\beta_1 C_A \cdot \beta_2 C_B}{\beta_1 C_A + \beta_2 C_B} \cdot \frac{A}{RT} \qquad\qquad (II, 49)$$

In this case, the rate of the reaction turns out to be proportional to the change of free energy. This could explain the proportionality between the reaction rate and the change of free energy which is sometimes found, e.g., in the kinetics of photographic development[25].

In the simultaneous presence in the same mixture of several reversible reactions bound by a common component with each reaction taking place in the diffusional region

in both the forward and the reverse direction, complete
equilibrium will establish itself at the surface, and the
ratio of the amounts of products formed will be entirely
governed by that equilibrium without depending on the rates
of the individual reactions.

Parallel and Consecutive Reactions

In many cases, several different products can be
formed from one and the same initial reactant. The reactions
leading to the formation of the different products can be
either parallel or consecutive.

An example is the combustion of carbon. The reaction between carbon and oxygen brings forth two products,
carbon dioxide and carbon monoxide. Until recently it was
commonly assumed that the formation of these products is the
consequence of two consecutive reactions

$$C + O_2 = CO_2$$

$$CO_2 + C = 2CO$$

Recently, Chukhanov and Grozdovskii[26] advanced a contrary
idea, according to which the formation of these two products
of the combustion is the result of two parallel reactions

$$C + xO_2 = yCO_2 + 2(x - y)CO$$

$$C + \frac{1}{2}O_2 = CO$$

As we have remarked[27], experimental data provide
no basis yet for such a conclusion.

With the reaction passing to the diffusional region, the proportions of the yields of the different products
will vary, the amount depending on whether the reactions are
parallel or consecutive.

In the case of parallel reactions, the amounts of products formed will be proportional to the rates of the corresponding reactions, and this relation will hold in both the kinetic and in the diffusional region. This is the only instance where even in the diffusional region, the true kinetics at the surface remains essential, and where some conclusions about this true kinetics can be drawn from experimental data obtained in the diffusional region.

The situation is entirely different in the case of consecutive reactions. Let us consider two consecutive reactions in which the initial reactant A gives the product B and the latter, in turn, is converted into the product C

$$A \longrightarrow B$$

$$B \longrightarrow C$$

We shall assume both reactions to be unimolecular and designate their rate constants by k_1 and k_2, respectively. The concentrations of the substances A, B, and C, in the volume will be designated by C_A, C_B, and C_C, and, at the surface, by C'_A, C'_B, and C'_C. The diffusion velocity constants will be designated by β_1, β_2, and β_3. The rate of consumption of the substance A will be designated by q_A, the rate of production of B by q_B, and the rate of production of C by q_C.

For the quasistationary state we have

$$q_A = \beta_1(C_A - C'_A) = k_1 C'_A$$

$$q_B = \beta_2(C'_B - C_B) = k_1 C'_A - k_2 C'_B \qquad \text{(II, 50)}$$

$$q_C = \beta_3(C'_C - C_C) = k_2 C'_B$$

From these equations we can find C'_A, C'_B, C'_C. We shall need only the values of C'_A and C'_B:

$$C'_A = \frac{\beta_1 C_A}{k_1 + \beta_1} \qquad (II, 51)$$

$$C'_B = \frac{k_1 \beta_1 C_A + (k_1 \beta_1 + \beta_1 \beta_2) C_B}{(k_1 + \beta_1)(k_2 + \beta_2)} \qquad (II, 52)$$

Let us consider the simplest and most interesting case when the concentration C_B of the substance B in the volume is zero. Then

$$C'_B = \frac{k_1 \beta_1 C_A}{(k_1 + \beta_1)(k_2 + \beta_2)} \qquad (II, 53)$$

The rates of formation of the substances B and C in this case will be

$$q_B = k_1 \frac{\beta_1 C_A}{k_1 + \beta_1} - k_2 \frac{k_1 \beta_1 C_A}{(k_1 + \beta_1)(k_2 + \beta_2)} = \frac{k_1 \beta_1 \beta_2 C_A}{(k_1 + \beta_1)(k_2 + \beta_2)}$$

$$q_C = \frac{k_1 k_2 \beta_1 C_A}{(k_1 + \beta_1)(k_2 + \beta_2)}$$

The ratio of the yields of the two products of the reaction will then be

$$\frac{q_B}{q_C} = \frac{\beta_2}{k_2} \qquad (II, 54)$$

This extremely simple result shows that the ratio of the yields of the two products depends only on whether the second reaction takes place in the kinetic or in the diffusional range.

If the second reaction lies in the kinetic range ($\beta_2 \gg k_2$), then $q_B \gg q_C$, i.e., practically only the substance B will be formed. If, on the contrary, the second reaction lies in the diffusional range ($\beta_2 \ll k_2$), then $q_B \ll q_C$, i.e., the sole product of the reaction will be the

substance C, even though it is in no way a primary product
but is formed exclusively from the substance B the concentration of which in the volume is zero.

This is entirely natural. The substance B,
formed at the surface, is removed therefrom at a rate equal
to the rate of diffusion. If the second reaction takes
place in the diffusional region, i.e., its rate is considerably greater than the rate of diffusion, the substance B
will have time to react and to get converted into C very
much faster than it can be removed from the surface through
diffusion.

Thus, with even the smallest concentration of the
substance B in the volume, we can get in the reaction
products the highest content of substance C; this does
not in any way imply that C is a primary reaction product.
Specifically, in the combustion of carbon at high temperatures when the reaction $CO_2 + C$ goes over into the diffusional range, we should observe CO in the reaction products; the fact that it is not the sole product of the combustion is a result of the secondary reaction of oxidation
of CO. The possibility of obtaining substantial amounts
of CO at the smallest concentration of CO_2 in the volume
(oxygen zone) does not by any means prove, as is sometimes
assumed, that CO is a primary product of the combustion.

This could be proved only if it were demonstrated
that the reaction CO + C does not lie in the diffusional
region. This, however, we are in no position to demonstrate
as we have no data on the rate of this reaction in the presence of oxygen.

A second interesting and practically important
instance of a process where two different products can be
obtained from the same initial substance is the catalytic
production of nitric acid from ammonia. Here, ammonia can
give either nitric oxide or elementary nitrogen, the latter
reaction being undesirable (defixation of nitrogen). Higher

temperature and shorter contact time favor the formation of NO and depress the yield of molecular nitrogen.

It is clear that in this case there are no consecutive reactions. If there were, higher temperature would increase the rate of the second reaction and, consequently, increase the yield of the final product N_2; actually, the contrary is observed. Moreover, direct experiments have shown that defixation of nitrogen is found at temperatures of the catalyst surface at which catalytic decomposition of NO does not yet take place at all. On the other hand, if we assume parallel reactions, we are unable to account for the effect of the contact time.

Apparently, an essential role is played in this case by unstable intermediate products formed in the oxidation of ammonia. The formation of elementary nitrogen must be explained by a reaction of NO with such intermediate products.

Autocatalytic Reactions

In the case of autocatalytic reactions, an increase of the diffusion velocity can result in removal of the catalyzing product from the surface and, consequently, in a decrease of the rate of the reaction. Let us consider the simplest case of an autocatalytic reaction of the first order with respect to the product formed. We designate the concentration of the product in the volume by C, and at the surface by C'. We limit ourselves to the case where the concentrations C and C' are small compared with the total stock of the reactant in the system. The concentration of the initial reactants can then be considered constant, and only the dependence of the rate of reaction on the concentration of the product formed need be taken into account:

$$f(C') = kC' \qquad\qquad (II, 55)$$

This expression is formally exactly identical to (II, 2a)

but differs from it in the meaning of the magnitude f(C').
In (II, 2a) it means the rate of consumption of the substance
present at the concentration C'; here it means the rate
of its formation.

For the quasistationary state, we get the equation

$$\beta(C' - C) = kC' \qquad (II, 56)$$

which differs from (II, 2) only in the sign.

Solving equation (II, 56) for C', we find

$$C' = \frac{\beta C}{\beta - k} \qquad (II, 57)$$

If the diffusion velocity is high, $\beta > k$, we observe a
quasistationary course of the process; with $\beta \gg k$, the
concentration of the reaction product at the surface will
tend to the concentration in the volume.

In contrast, at low diffusion velocities, $\beta < k$,
a quasistationary course in the adopted approximation becomes
impossible. Even at the smallest concentration of the reaction product in the volume its concentration at the surface
will grow without limit until the stock of reactant is exhausted, and then our assumption of the independence of the
rate of the reaction of the concentrations of the initial
reactants will become inapplicable. At $\beta = k$, we shall
observe the critical condition of the transition from the
stationary to the nonstationary course of the reaction, entirely analogous to the critical conditions of chain ignition
discussed in the chain theory.

Thus, at low diffusion velocities we can observe
a high rate of reaction as a result of accumulation at the
surface of large amounts of the catalyzing product. An increase of the velocity of diffusion (e.g., through intensified agitation) can cause sharp decrease of the rate of reaction as a result of transition to a stationary course.

Experimentally, a decrease of the rate of reaction with increasing rate of stirring was observed in the dissolution of copper in nitric acid[28]. This reaction represents a typical autocatalytic process because it is accelerated by the nitrite formed in the reaction.

Uniformly Accessible Surface

The simple relations just discussed are valid in the simple case when the reaction, which involves only one reactant diffusing from the volume to the surface, takes place at a surface all points of which are equally accessible with respect to diffusion (uniformly accessible surface). Obviously, in order for the above considerations to be applicable, it is necessary that the reactant be diluted by a large excess of an extraneous inert gas (or diluent) or of the reaction products. With only the pure reactant gas present in the volume, the diffusional resistance is zero. We have disregarded some relatively secondary factors, such as the mass flow arising as a result of the change of volume in the reaction. These factors will be discussed further in Chapter III.

The most serious limitation of the foregoing simple theory is the assumption of a uniformly accessible surface which is true only in the very simplest cases. If the reaction takes place at the surface of a body of any geometric shape without deep cracks or pores and the body is placed in a large vessel, such a surface can be considered uniformly accessible without serious error. The same will be true if the reaction takes place at the walls of a channel along which the gas or liquid flows. However, the results obtained for a uniformly accessible surface are entirely inapplicable to reactions taking place on porous or powdery materials.

Porous Surface

The rate of reaction on a porous or powdery material is composed of the rates on different portions of the surface, characterized by different accessibilities with respect to diffusion. The total rate of reaction depends both on the shape and diameter of the pores and on the thickness and geometric shape of the layer of the material.

For a finely porous material one can, following Zeldovich[8*], view the problem as if the reaction were taking place over the whole volume occupied by the material.

The results will be immediately applicable also to cases, such as the dissolution of a gas in a liquid, which are accompanied by a chemical reaction in the volume of the liquid phase. In other words, we shall view the process as if the reaction were taking place homogeneously, but with the reactant supplied by diffusion from the other phase.

We shall consider the limiting case where the layer of material is sufficiently thick to be considered infinitely thick.

In order to analyze the problem independently of the shape and diameter of the pores, we shall describe the diffusion within the mass of the porous material with the aid of the effective diffusion coefficient D', so defined that the equation of diffusion in the mass of the material is of the form

$$\frac{\partial \zeta}{\partial t} = D' \Delta \zeta - f'(\zeta) \qquad\qquad (\text{II, } 58)$$

where ζ is the concentration of the reactant at some point within the mass of the porous material, and $f'(\zeta)$ is the effective rate of reaction, i.e., the amount of substance consumed at that point per unit time in unit volume as a result of the reaction.

According to our assumption of an infinite thick-

---

*See also: Thiele, E. W., Ind. Eng. Chem., 31 916 (1939).

ness of the layer of porous material, we shall consider its
surface plane. We shall then have to consider diffusion
only in one direction perpendicular to the surface and the
conditions of the problem will contain no magnitudes of the
dimension of a length.

In this case, equation (II, 58) will be written
in the form

$$\frac{\partial \zeta}{\partial t} = D' \frac{\partial^2 \zeta}{\partial x^2} - f'(\zeta) \qquad\qquad (II, 59)$$

where x is the distance of the point under consideration
from the surface.

After the lapse of some time, a stationary concentration distribution will have established itself in the
mass of the material, at which

$$\frac{\partial \zeta}{\partial t}$$

will be zero. This stationary state is examined in the
theory of Zeldovich. For the stationary state, equation
(II, 59) takes the form

$$D' \frac{d^2 \zeta}{dx^2} = f'(\zeta) \qquad\qquad (II, 60)$$

In order to be able to use the dimensional method, we shall
assume that the true kinetics of the reaction is of an nth
order

$$f'(\zeta) = k' \zeta^n \qquad\qquad (II, 61)$$

Equation (II, 60) will now take the form

$$D' \frac{d^2 \zeta}{dx^2} = k' \zeta^n \qquad\qquad (II, 62)$$

Let the concentration of the reactant at the surface be C;
we shall take this magnitude as the natural yardstick of
concentration. No yardstick of length is necessary as there
are no magnitudes of the dimension of a length in the conditions of the problem.

With the dimensionless concentration $\zeta' = \frac{\zeta}{C}$,
equation (II, 62) becomes

$$D'C \frac{d^2\zeta'}{dx^2} = k'C^n\zeta'^n \qquad (II, 63)$$

or

$$\frac{D'}{k'C^{n-1}} \frac{d^2\zeta'}{dx^2} = \zeta'^n \qquad (II, 64)$$

It is obvious that the equation in terms of dimensionless
variables will take the simplest form if one introduces the
dimensionless coordinate

$$\xi = x \sqrt{\frac{k'C^{n-1}}{D'}} \qquad (II, 65)$$

Equation (II, 64) will then go over into

$$\frac{d^2\zeta'}{d\xi^2} = \zeta'^n \qquad (II, 66)$$

with the boundary conditions

$$\zeta' = 1 \text{ at } \xi = 0$$

$$\zeta' = 0 \text{ and } \frac{d\zeta'}{d\xi} = 0 \text{ at } \xi = \infty$$

Equation (II, 66) is easy to integrate analytically. But
we can obtain the main result immediately with the aid of
the dimension method. We are interested in the total amount
of substance reacting per unit time and unit free surface

area of the layer (not the total internal surface area of the pores) which we shall designate by $\left(\frac{dm}{dt}\right)$. It is equal to the diffusion flow passing through the free surface of the layer

$$\left(\frac{dm}{dt}\right) = -D'\left(\frac{d\zeta}{dx}\right)_{x=0} \qquad \text{(II, 67)}$$

The magnitude

$$\left(\frac{d\zeta}{dx}\right)_{x=0}$$

is found by solving equation (II, 66). As neither equation (II, 66) nor the boundary conditions contain any parameters, its solution cannot contain any parameter either and must be of the form $\zeta' = f(\xi)$. Consequently,

$$\left(\frac{d\zeta'}{d\xi}\right)_{\xi=0}$$

will be a constant number.

Obviously, with the adopted direction of the x axis,

$$\left(\frac{d\zeta'}{d\xi}\right)_{\xi=0}$$

is negative; we therefore shall designate it by $-A$.

Substituting for ζ' and ξ their expressions by ζ and x, we get

$$\frac{d\zeta}{dx} = \sqrt{\frac{k'c^{n+1}}{D'}}\,\frac{d\zeta'}{d\xi}$$

Substituting this in (II, 67), we get

$$\left(\frac{dm}{dt}\right) = -\sqrt{D'k'c^{n+1}}\left(\frac{d\zeta'}{d\xi}\right)_{\xi=0} = A\sqrt{D'k'c^{n+1}}$$

$$\text{(II, 68)}$$

Thus, in the case under consideration, the rate of the overall process is proportional to the geometric mean of the
rate of reaction $k'C^n$ and the diffusion velocity $D'C$.

Analytical solution of equation (II, 66) gives
the value of the proportionality factor A. Integration,
with the boundary conditions taken into account, gives

$$\frac{d\zeta'}{d\xi} = -\sqrt{\frac{2}{n+1}}\; \zeta'^{\frac{n+1}{2}}; \quad \zeta' = \left(\frac{n-1}{\sqrt{2(n+1)}}\,\xi + 1\right)^{\frac{2}{1-n}}$$

We remark that at $n > 1$ the concentration remains finite at any distance from the surface and only asymptotically tends to zero. On the other hand, for a fractional-
order reaction $(n < 1)$ the reactant concentration vanishes
at a finite distance from the surface, namely at the distance

$$\xi = \frac{\sqrt{2(1+n)}}{1-n}$$

At $n = \frac{1}{3}$, the concentration becomes zero at the distance
$\xi = 2.45$; at $n = \frac{2}{3}$, at the distance $\xi = 5.50$ from the
surface.

Thus, the characteristic feature of a fractional-
order reaction is the final depth of penetration of the reaction into the layer of material, whereas for $n > 1$ this
depth is essentially infinite, and the effective depth of
penetration of the reaction corresponds only to the distance
at which there is a significant fall of the concentration.

As we have pointed out in the Introduction, the
case of a fractional-order reaction is in no way unreal.
For heterogeneous reactions the order is very often less
than unity.

In the particular case $n = 1$ (first-order reaction) the expression for ξ' becomes indefinite, and when
lifted gives

$$\zeta' = e^{-\sqrt{\frac{2}{n+1}}\ \xi}$$

For any n one gets for the proportionality factor

$$A = -\left(\frac{\partial \zeta'}{\partial \xi}\right)_{\xi=0} = \sqrt{\frac{2}{n+1}}$$

(according to the boundary condition, at $\xi = 0$, $\xi' = 1$).

For the macroscopic rate of reaction we get, finally

$$\left(\frac{dm}{dt}\right) = \sqrt{\frac{2}{n+1}\ D'k'C^{n+1}} \qquad\qquad (II, 69)$$

where n is the order of the reaction.

We now can formulate accurately the conditions under which it is permissible to use the assumption of infinite thickness of the layer of the material. By formula (II, 65) the natural yardstick of length (obtained from the equation itself) is in this case the magnitude

$$\sqrt{\frac{D'}{k'c^{n-1}}}$$

if the thickness of the porous material is great compared with this magnitude, the assumption is permissible. Thus, the greater is the rate of reaction, the smaller the thickness of the layer at which the assumption is usable. Conversely, at a finite thickness of the layer, decrease of the rate of reaction leads to a region where the thickness of the layer becomes very much smaller than

$$\sqrt{\frac{D'}{k'c^{n-1}}}$$

here, the whole inner surface of the material will behave

entirely as a free surface, i.e., we shall find ourselves in a pure kinetic region.

For the sake of instructiveness we can introduce the notion of effective depth of penetration of the reaction into the layer of the porous material L.

Except for a dimensionless factor of the order of unity, one can define the depth of penetration of the reaction as

$$L \approx \sqrt{\frac{D'}{k' C^{n-1}}} \qquad\qquad (II, 70)$$

In particular, for a first-order reaction

$$L \approx \sqrt{\frac{D'}{k'}} \qquad\qquad (II, 70a)$$

We see that the greater the rate of the reaction the smaller the depth of its penetration into the layer of the material.

If the total thickness of the layer H is great compared with the depth of penetration L, one can make use of the calculation just considered, which rests on the assumption of an infinite thickness of the layer. If L becomes of the order of H, the macroscopic rate of reaction will depend on the ratio $\frac{L}{H}$. Finally, if the rate of reaction becomes sufficiently fast for the depth of penetration L to fall to values comparable with the diameter of individual pores h, the whole approach of Zeldovich becomes unsuitable. It will neither be possible to introduce an effective diffusion coefficient nor an effective rate constant per unit volume, as the reaction will take place only at the outer surface of the lump. Further increase of the rate of reaction will no longer result in a decrease of the depth of its penetration into the interior of the lump, as this depth is of the order of the diameter of individual pores, or smaller. Consequently, at

$$L \gtrless h$$

one can disregard completely the effect of the porosity and consider that the reaction takes place only at the outer surface of the lump.

If the assumption of infinite thickness of the layer is applicable, the overall rate of the reaction is described by formula (II, 69). The temperature dependence of the rate of reaction should correspond to half the activation energy. With the aid of the notion of effective depth of penetration into the layer of material, we can give a very instructive interpretation to formula (II, 69). With increasing true rate of the reaction, the depth of penetration decreases; according to (II, 70), it is inversely proportional to the square root of the rate of reaction. On the other hand, the macroscopic rate of the reaction is proportional to the product of the true rate of the reaction and to the depth of penetration. Consequently, the macroscopic rate of the reaction increases proportionally to the square root of the true rate of reaction.

In (II, 69) and the preceding formulas, we considered the concentration C of the reactant directly at the free surface of the layer of porous material as given, and we took into account only the diffusion within that layer. If the reaction is fast, diffusion in the gaseous (or liquid) phase to the surface of the layer can also become significant. It can be easily taken into account by methods already known to us. The diffusion flow from the volume to the surface is

$$q = \frac{Nu\ D}{d} (C_o - C)$$

where C_o is the concentration of reactant in the volume; C, at the surface of the layer; and D is the true diffusion coefficient of the reactant in the volume.

Equating the diffusion flow to the right-hand

member of the expression (II, 69), we get the algebraic
equation

$$\frac{Nu\ D}{d}(C_0 - C) = \sqrt{\frac{2}{n+1}\ D'k'C^{n+1}} \qquad (II,\ 71)$$

for the determination of the quasistationary concentration
C of the reactant at the surface. In the particular case
when the true kinetics of the reaction is of the first order,
equation (II, 71) takes the form

$$\frac{Nu\ D}{d}(C_0 - C) = \sqrt{D'k'}C$$

hence

$$C = \frac{\dfrac{Nu\ D}{d}C_0}{\dfrac{Nu\ D}{d} + \sqrt{D'k'}}$$

and

$$\left(\frac{dm}{dt}\right) = \frac{\dfrac{Nu\ D}{d}\sqrt{D'k'}}{\dfrac{Nu\ D}{d} + \sqrt{D'k'}}C_0$$

In this problem, there can be four limiting regions:

1. At $\sqrt{D'k'} \gg \frac{Nu\ D}{d}$ the rate of the overall
process is determined by diffusion in the volume

$$\left(\frac{dm}{dt}\right) = \frac{Nu\ D}{d}C_0$$

The concentration of the reactant even at the
surface of the layer, and even more in the interior of the pores, is considerably smaller than
in the volume

$$C \ll C_0$$

Following a proposal of Vulis[9], this region is

termed the "outer diffusional region".

2.   At $\frac{Nu\,D}{d} \gg \sqrt{D'k'}$ and $H \gg L \gg h$ (where
L is the depth of penetration of the reaction
into the interior of the lump, defined by formula
(II, 70); H is the thickness of the lump or of
the layer of porous material; and h is the mean
diameter of single pores). The determining stage
is diffusion in pores. The concentration of the
reactant at the free surface of the layer is very
close to its concentration in the volume, $C \approx C_o$,
whereas the concentration in the pores falls practically to zero. The rate of the reaction in this
region is expressed by formula (II, 69). This
region could be termed the region of diffusion in
pores or, following a very pertinent proposal of
Vulis, the "inner diffusional region".

3.   At $\frac{Nu\,D}{d} \gg \sqrt{D'k'}$ and $L \gg H$, the macroscopic kinetics coincides with the true kinetics
at the surface. In this "inner kinetic region",
the concentration of the reactant in the pores
throughout the thickness of the layer coincides
with the concentration in the volume. In the
inner kinetic region, the whole inner surface of
the porous material is in operation.

4.   Finally, at $L \lessgtr h$ and $\frac{Nu\,D}{d} \gg k$ (where
k is the true rate constant of the reaction at
the surface) we observe the "outer kinetic region"
where the macroscopic kinetics also coincides with
the true kinetics at the surface, but the reaction
takes place only at the outer surface of the lumps
of the porous material.

Evidently, in the inner kinetic region the macroscopic rate of reaction is proportional to the volume of the
porous material, and in the outer kinetic region it is proportional to its surface area.

In the region intermediate between the outer and the inner diffusional regions, the concentration field and the rate of the overall process can be obtained by solving equation (II, 71). For the simplest case of a monomolecular reaction this was done above. The corresponding formulas for the region intermediate between the inner diffusional and the inner kinetic regions can be obtained by solving equation (II, 58) under the changed boundary conditions,

$$\frac{d\zeta}{dx} = 0 \quad \text{at} \quad x = L$$

With these boundary conditions, the equation can be integrated by elementary functions only for the simplest case of a monomolecular reaction, but even in this simplest case the final formulas are quite involved and will not be given here.

It still remains to clarify the significance of the effective diffusion coefficient D' and the effective rate of reaction $f'(\zeta)$ or the effective rate constant k'. These magnitudes have a very simple meaning when one considers not a reaction in a porous material but a homogeneous reaction in which the reactant is supplied by diffusion from another phase; for example, the absorption of a gas by a liquid accompanied by a chemical reaction in the liquid. If convection in the liquid is taken to be absent, this process will be described by formula (II, 71), with D' designating simply the diffusion coefficient of the reactant in the liquid, and k' the rate of the homogeneous chemical reaction in the liquid.

In the case of a porous or a powdery material, the meaning of these magnitudes will be somewhat different. It is related to the structure of the porous surface. Let us take the simplest model of a porous material and consider the pores as capillaries running from the free surface, without breaks or intersections throughout the whole thickness of the layer. As characteristics of the pores, we introduce

the mean pore diameter h, the number N of pores per unit
surface area, and the "labyrinth coefficient" X which
is defined as the mean distance along the pores corresponding
to unit lengths in the direction perpendicular to the surface

$$X = \frac{dl}{dx}$$

l is the distance measured along the direction of the pores,
and x is the distance measured perpendicularly to the surface.

The surface area of the pores per unit volume of
the layer is equal

$$X \, N\pi h \; .$$

Consequently, the effective rate of reaction $f'(\zeta)$ per
unit volume is related with the true rate of reaction $f(\zeta)$
per unit reacting surface area by the expressions

$$f'(\zeta) = X \, N\pi h f(\zeta) \qquad\qquad (II, \, 72)$$

$$k' = X N\pi h k \qquad\qquad (II, \, 72a)$$

where the rate constant k is defined as in (II, 2a).

Fick's law of diffusion in the pores will be
written

$$q^* = D \, \frac{dC}{dt}$$

where q^* is the diffusion flow per unit surface area of
the free cross section of the pores. In order to find the
effective diffusion coefficient D', it is necessary to
find the diffusion flow per unit area of the total cross
section of the layer, relative not to $\frac{dC}{dl}$ but to $\frac{dC}{dx}$.
Bearing in mind that

$$\frac{dC}{dl} = \frac{dC}{dx} \frac{dx}{dl} = \frac{1}{X} \frac{dC}{dl}$$

and that the area of the free cross section of the pores per unit area of the total cross section of the layer is

$$\Omega = N \frac{\pi h^2}{4}$$

we get for the diffusion flow q per unit surface area of the total cross section of the layer

$$q = N \frac{\pi h^2}{4} q^* = DN \frac{\pi h^2}{4} \frac{dC}{dl} = \frac{dN}{x} \frac{\pi h^2}{4} \frac{dC}{dx}$$

$$D' = D \frac{N}{x} \frac{\pi h^2}{4} \qquad (II, 73)$$

Substituting (II, 72a) and (II, 73) in (II, 69) we get

$$\left(\frac{dm}{dt}\right) = \sqrt{\frac{2}{n+1} N^2 \frac{\pi^2 h^3}{4} DkC^{n+1}} \qquad (II, 74)$$

One can pass, for the characterization of the pores, to a directly measurable quantity, the porosity of the material. This quantity is defined as the ratio of the pore volume and the total volume of the layer. In our model of a porous layer, the porosity is equal to the ratio of the area of the free cross section of the pores and the area of the total cross section of the layer. This ratio we have designated by Ω. Expressing N by this magnitude, we can represent (II, 73) in the form

$$\left(\frac{dm}{dt}\right) = \Omega \sqrt{\frac{8}{n+1} \frac{Dk}{h} C^{n+1}} \qquad (II, 75)$$

This formula is particularly instructive. The magnitude Ω is dimensionless. Formula (II, 75) shows the reactivity of the porous material in the range of diffusion in the pores to be directly proportional to the porosity at constant pore diameter. With increasing pore diameter at constant porosity,

the reactivity decreases as the square root of the pore
diameter owing to the decrease of the total surface
area.

        In practical calculations it is most important to
decide in which of the above characterized four limiting
regions the process takes place in the given concrete instance. In order to find an answer to this question, it is
necessary to calculate the depth of penetration L of the
reaction into the lump by formula (II, 70) or (II, 70a).
The effective reaction rate constant k' per unit volume
of the porous material can be determined directly from experimental data of the rate of reaction at very low temperatures where the reaction is known to take place in the
inner kinetic region. The effective diffusion coefficient
can be estimated with the aid of formula (II, 73) if the
pores are wide enough and the pressure high enough for the
pore diameter to be large compared with the mean free path
of the molecules. With small pore diameters and at low
pressures, we can find ourselves in the Knudsen region where
Fick's law is inapplicable. In order to use formula (II, 73),
it is necessary to know the mean pore diameter h and the
number N of pores per unit area. To calculate these two
unknowns one can utilize any other two magnitudes which are
related therewith and which are easily accessible to experimental determination, such as gas permeability and total
surface area (from adsorption measurements), or gas permeability and porosity (from the ratio of specific and volume
weight). Such calculations have been made by Roiter[10] for
ammonia catalysts.

        In cases where L, calculated by formula (II, 70),
is large compared with the dimensions of the catalyst grain,
one can assume that the whole inner surface of the grain is
in operation. In cases where the magnitude L is comparable
with the pore diameter, only the outer surface of the grain
is in operation. If L is large compared with the pore
diameter but small compared with the dimensions of the grain,

diffusion in pores is determining, and the process takes it course according to Zeldovich's formula (II, 69). In each case it is, moreover, necessary to ascertain to which extent diffusion in the volume can be significant.

The above considerations are contingent on a number of simplifying assumptions: only one diffusing substance is considered, the Stephan flow (cf. Chapter III) is disregarded, and possible temperature changes are not taken into account. An analysis of the permissibility of these simplications can be found in the recently published paper of Pshenitskii and Rubinshtein[28].

Diffusion Across Membranes

In living organisms, diffusion processes across colloidal films or membranes play a great role. In some instances diffusion across high-polymer films is important in technology.

Diffusion across membranes is not simply a physical process. It is closely tied with sorption and with nonequilibrium chemical processes. The simplest theory of diffusion across membranes is the theory of equilibrium sorption. According to it, the diffusion process is linked with sorption of the diffusing substance, i.e., with its dissolution in the material of the film. The substance then diffuses in the dissolved state, after which desorption takes place at the other side.

This simplest theory rests on the assumption of equilibrium sorption, i.e., it is assumed that the sorption equilibrium establishes itself extremely rapidly. In this case the amount of substance diffused should be proportional to the time, i.e., the process should formally be described by Fick's law.

Experimental data show that in reality the process follows a nonstationary course with time. At small concentrations of the diffusing substance or for weakly sorbed

substances, the velocity of the diffusion first increases
with time and then becomes constant. In the initial stages
there is, apparently, gradual penetration of the diffusing
substance into the thickness of the film until a stationary
concentration distribution is established. Such a pattern
was found by several investigators in such processes as the
diffusion of hydrogen[18] or water vapor[19] across polymer
films. In such instances diffusion is determined only by
physical sorption processes without any deep chemical interaction between the adsorbed substance and the material
of the film. This form of the process will be called passive diffusion.

Entirely different phenomena can be observed at
high concentrations of substances strongly sorbed by the
material of the film. Here we have activated diffusion,
i.e., a slow process wherein diffusion is accompanied by
active chemical interaction between the diffusing substance
and the material of the films. These processes can lead
either to fixation of the diffusing substance within the
film with strong binding (negative active diffusion) or to
a separation of the initially bound substance (positive
active diffusion).

In cases of active diffusion the film, although
it is permeable to the given substance and impermeable to
other (inert) substances, differs markedly in its behavior
from the classic (equilibrium) semipermeable septum.

The concentration of the diffusing substance on
both sides of the membrane is not equalized, no matter how
much the diffusion is prolonged. This is clearly demonstrated
in studies of diffusion by the manometric method.

In the case of negative active diffusion, the
concentration behind the membrane is at all times lower than
the concentration in front of it. The diffusion process is
not activated but is slowed down as a result of the saturation of the membrane by the diffusing substance. Ultimately,

in a manometric investigation of diffusion, one can observe
a fall of the pressure behind the membrane.

The most natural explanation of the negative active diffusion seems to be supplied by the theory of two-
stage nonequilibrium sorption. According to it, the primary
process of rapid (equilibrium) sorption of the diffusing
substance by the material of the film is followed by the
much slower process of passage of the sorbed substance into
a more strongly bound state. It is possible that, as is
assumed in the colloid-chemical theories of sorption, the
first stage corresponds to the solution of the diffusing sub-
tance in the intermicellar liquid and the second to its penetration into the interior of the colloidal micelles. Swell-
ing of the micelles decreases the intermicellar interstices
and thus counteracts diffusion. After complete saturation
of the film by the diffusing substance, its permeability
decreases.

Diametrically opposed to this behavior are phenomena of positive active diffusion, where the concentration
of the diffusing substance behind the membrane is found
higher than the concentration of the diffusing substance in
front of it. This form of active diffusion, which apparently
is absent in nonliving nature but is widespread in living
organisms, is called "secretion" in physiology. Here, the
transport of matter from lower to higher concentration takes
place at the expense of simultaneously occurring nonequilibrium chemical processes.

Secretion phenomena are among the most important
manifestations of the life activity of a living organism.
They take place in all glands and in such organs as the
kidneys. There exists a theory of the respiratory process
in which it is assumed that oxygen is secreted in the lungs.
However, in the opinion prevailing in present-day physiology,
this theory is wrong and the transport of oxygen in the lung
can be interpreted entirely on the basis of the usual passive
diffusion.

A most striking instance of secretion of gases[20] is provided by the phenomena taking place in the floating bladders of deep-water fish. The fish equilibrates the pressure of the surrounding water by producing in its bladder an equal pressure of gas. In deep-water fish, this gas consists mainly of oxygen, and its pressure can attain several hundreds of atmospheres. But the fish gets its oxygen from the water where it is present at a concentration no higher than that which corresponds to equilibrium with atmospheric air. The oxygen obtained from the water is absorbed by the hemoglobin of the blood, and then a special oxygen gland brings about its secretion to the floating bladder.

Thus, this oxygen gland fulfills the function of a powerful compressor which compresses the oxygen from a partial pressure of some 0.21 atmosphere under which it is present in the air to several hundreds of atmospheres. The living organism thus appears to be much more efficient than human technology. The task, which in technology requires massive and heavy compressors, is performed in the body of the fish by a small, delicate gland.

As the passage of a substance from a lower to a higher concentration would be in sharp conflict with thermodynamics, it is clear that phenomena of positive active diffusion must also be linked with nonequilibrium chemical processes. But, in contrast to the case of negative active diffusion, these processes ought to lead not to a binding of diffusing substance but, on the contrary, to its separation from the initially bound state. They ought to occur not over the whole thickness of the membrane but at the side where the diffusing substance is liberated and separated. Thus, a membrane capable of secretion must have a sufficient thickness, and chemical conditions must not be the same over the whole thickness of the membrane.

If one could make artificial membranes capable of positive active diffusion, i.e., the artificial analog of secretory action, it would be a great step forward on the road of artificial imitation of biological processes.

So far, no such artificial secretion has been observed. Phenomena of negative active diffusion are, in a way, akin to secretion phenomena, but in the opposite sense. One can expect that a more detailed study thereof may bring us closer to an understanding of the mechanism of biological diffusion.

Diffusion Through Pores

When we spoke of diffusion across membranes, we referred to membranes without porosity. Penetration of a substance through such a membrane is possible only by way of its dissolution or, generally, its sorption by the substance of the membrane.

In contrast, in the case of porous partitions, there can be purely physical diffusion of the gas through the pores. This process takes place according to the usual laws of gaseous diffusion. Diffusion through pores can be easily distinguished from sorptive diffusion by comparing the diffusion of different gases through the same membrane. In the case of diffusion through pores, gases of low molecular weight (as hydrogen) should diffuse more easily than those of high molecular weight. When sorption is a major process, the rate of diffusion varies directly with the strength of sorption in the film.

Experimental data show that for colloidal membranes porosity usually has a secondary significance and the mechanism of the diffusion is in the main bound with sorption.

Formation of Solid Films

An interesting and widespread instance of diffusional kinetics is a reaction in which the reaction product forms at the surface a solid film through which the reacting gas is bound to diffuse. An important example is the oxi-

dation of metals by the oxygen of the air[21] resulting in a
solid oxide film through which the oxygen must diffuse.

Simplest cases of this type can be described by
a very elementary theory. Let us designate the thickness
of the film by δ. The velocity of diffusion across the
film is inversely proportional to its thickness, and the
rate of growth of the film is proportional to amount of gas
passing through it.

Consequently, we have

$$\frac{d\delta}{dt} = \frac{A}{\delta} \qquad\qquad (II, 76)$$

where A is a constant proportional to the product of the
diffusion coefficient and the rate of reaction.

$$\delta = \sqrt{2\ At} \qquad\qquad (II, 77)$$

Thus, the thickness of the film increases proportionally to
the square root of the time. The amount of diffusing gas
and the rate of reaction are inversely proportional to the
thickness of the film, i.e., inversely proportional to the
square root of the time.

This result becomes valid after the lapse of a
sufficiently long time from the beginning of the process,
when a stationary concentration distribution of the diffusing
substance has established itself in the film. In the initial
period, the rate of reaction will depend on the time according to a different, more complex law. This problem has been
examined mathematically by Vulis[22] in his work on the theory
of combustion of ash containing coal.

From the point of view of the mathematical theory,
combustion of coal is entirely analogous to the processes
just referred to. Here, too, as the reaction progresses, a
solid film is formed through which the reacting gas must
diffuse. The role of the solid product is played by the
ash remaining after the combustion.

Vulis' work gives an exact solution of the diffusion equation in this instance and formulas for the time
dependence of the thickness of the film and of the rate of
reaction. At the limit, after a sufficient time has elapsed
from the beginning of the process, these formulas lead to a
rate inversely proportional to the square root of the time,
in agreement with the result obtained here in an elementary
way.

These simple relations are well confirmed by experiments for a number of important cases, in particular
for the action of iodine on silver and for the oxidation
by the oxygen of the air of a number of metals such as tung-
sten, copper, brass, iron, nickel. However, in a number of
other cases, the growth of the film is known to decrease
more rapidly than inversely proportionally to the square
root of the time, to follow a more nearly exponential law,
and to come to a practical halt after reaching a certain
thickness of film. Such phenomena are observed in many
cases of the action of oxidants on the surface of metals,
and lead to the formation corrosion-inhibiting protective
layers. A typical example is the formation of an oxide film
on the surface of aluminum.

The causes of the cessation of the growth of the
film are not clear at the present time. There exist several
theories, none of which can be considered sufficiently substantiated. One should point to the sharp quantum-mechanical
theory of Mott[23]; according to this theory, the diffusion
of the reactant across the film can be of the nature of a
tunnel transition under the potential barrier. The rate of
this process should decrease extremely rapidly with increasing thickness of the film, and this provides a likely explanation of the phenomena observed in the formation of protective films.

Microheterogeneous Processes

Interesting and important instances of diffusional

kinetics are encountered in the field of microheterogeneous reactions, i.e., of reactions taking place at the surface of disperse particles suspended in another phase. Examples of microheterogeneous processes are: combustion of coal dust blasted into the furnace by a stream of air; enzymatic reactions, taking place at the surface of the colloidal enzyme particles, and the analogous cases of catalysis by colloidal metals which Bredig has termed "inorganic ferments"; and hydrogenation of liquid oils under the catalytic action of a disperse catalyst suspended in the oil.

As a result of the small size of the particles, diffusion in microheterogeneous systems is strongly intensified. From the expression for the diffusion velocity constant,

$$\beta = \frac{Nu\ D}{d}$$

it follows that in cases where the influence of convection is of little importance, the velocity of diffusion is inversely proportional to the determining dimension.

The influence of convection on the diffusion process in microheterogeneous systems is weakened as the particles move along with the flow, and the velocity of the flow relative to the surface at which the reaction takes place is very much smaller than the absolute velocity of the flow. At the present time, however, we have no concrete data to permit an estimation of the intensity of the diffusion to the particles suspended in the flow and of the extent to which the influence of turbulence on the diffusion velocity is preserved in microheterogeneous systems.

Some theoretical considerations on that subject will be given in Chapter V.

Reactions in microheterogeneous systems can be subdivided into two groups. The first group shall include processes in which there is only diffusion of the reactant

to the surface of the disperse particles; the second, processes in which one of the reacting substances is supplied
by diffusion from another phase, and where not only its
diffusion to the surface of each individual particle but
also diffusion in the volume is essential.

In the case of processes of the first group, the
reactant is in the homogeneous solution from which it diffuses to the surface of each individual disperse particle.
The surface in this case can be considered uniformly accessible, and all relations discussed in the foregoing for a
uniformly accessible surface will hold. There will be, as
usual, a diffusional and a kinetic region. Only the value
of the diffusion velocity constant and its dependence on
velocity of the flow (or of agitation) will be different
from those for a continuous surface. In the present state
of our knowledge, the calculation of the diffusion velocity
constant is different.

Very peculiar is the case of a microheterogeneous
process in which one of the reactants is supplied by diffusion from the other phase. The best studied instance of
this type of process is the hydrogenation of unsaturated
compounds in the liquid phase with the aid of a catalyst
dispersed in the mass.

From the point of view of diffusional kinetics,
the hydrogenation processes were studied by Davis and his
collaborators[29] and in particular detail by Zhabrova and
Goldanskii[30].

It is natural to expect that at sufficiently high
rate of reaction at the surface, the rate of the process
should be governed by the diffusion of the hydrogen from
the gas phase; the diffusion of the unsaturated compound
from the medium immediately surrounding the catalyst particle
being much less difficult than the diffusion of the hydrogen
from the other phase. Actually, it was found experimentally
that in many cases the rate of the hydrogenation is inde-

pendent of the concentration of the hydrogenated compound
and proportional to the pressure of hydrogen. Such a dependence of the rate of the process on the concentrations
of the reactants is clearly in agreement with the assumption
that the process is limited by diffusion of the hydrogen.
Deviations from the proportionality between the rate of the
hydrogenation and the concentration of the hydrogen were
observed only near the critical point, where the solubility
of the hydrogen in the mass no more obeys Henry's law.

In processes of this kind it is necessary to take
into account, on the one hand, the diffusion of the reactant
(in the case of hydrogenation, of hydrogen) in the volume
of the liquid phase, and, on the other hand, its diffusion
to the surface of each individual particle of catalyst. Let
us consider the simplest case where the kinetics of the reaction at the surface of the disperse particles is of the
first order. We can then, for each individual disperse particle (the surface of which can, of course, be considered
uniformly accessible) make use of formula (II, 7) according
to which the effective rate constant per unit surface area
of the disperse particle is expressed by

$$k^* = \frac{k\,\beta}{k + \beta}$$

where k is the true rate constant of the chemical reaction
at the surface of the disperse particle, and β the velocity
constant of the diffusion to that surface

$$\beta = \frac{Nu\,D}{z}$$

where z is the mean diameter of the disperse particles.

If, on the basis of the above, one considers the
influence of convection on the diffusion to the surface of
the disperse particles unessential and if one treats the
disperse particles approximately as spheres of diameter z,
suspended in a limitless immobile medium, the value of the

Nusselt criterion will be 2 and we shall have for the diffusion velocity constant

$$\beta = \frac{2D}{z} \qquad\qquad (II, 78)$$

From the expression (II, 7) where the effective rate constant k* is referred to unit surface area of the disperse particle, we pass to the effective rate constant k', referred to unit volume of the liquid phase. To this end it is evidently necessary to multiply k* by the mean surface area σ of the disperse particle and by the number of disperse particles N in unit volume of liquid

$$k' = N\sigma k^* = \frac{N\sigma k\beta}{k + \beta} \qquad\qquad (II, 79)$$

The rate of reaction referred to unit volume of liquid will be expressed by k'C, where C is the concentration of the dissolved reactant in the liquid.

Let C_o designate the concentration of reactant at the surface from which the diffusion originates. It is related with the concentration in the gas by Henry's law. In order to express the concentration C by C_o, it is necessary to consider the diffusion process in the volume of the liquid phase. This can be done by two methods, corresponding to two physically possible limiting cases.

With a sufficiently intense agitation, one can consider the concentration C the same over the whole volume, with the exception of that in the effective film in which all the diffusion resistance is concentrated. The amount of reactant supplied by diffusion from the other phase will then be expressed as

$$\frac{D}{\delta} S(C_o - C)$$

where D is the diffusion coefficient, δ the effective film thickness, S the interface area through which the reactant diffuses.

In the stationary state this should be equal to the amount of substance consumed as a result of the reaction in the volume

$$k'\omega C$$

where ω is the volume of the liquid phase in which the reaction takes place.

Hence we obtain the equation for the determination of the quasistationary concentration of the reactant in the volume of the liquid phase

$$\frac{D}{\delta} S(C_0 - C) = k'\omega C \qquad (II, 80)$$

whence the concentration C will be expressed as

$$C = \frac{\frac{D}{\delta} S}{\frac{D}{\delta} S + k'\omega} C_0 \qquad (II, 81)$$

and the rate of the reaction per unit volume of the liquid phase as

$$-\frac{dc}{dt} = k'C = \frac{k' \frac{D}{\delta} S}{\frac{D}{\delta} S + k'\omega} C_0 \qquad (II, 82)$$

Substituting here k' from (II, 79), we get the final expression for the quasistationary rate of reaction. It is easy to remark that, in general, the rate of reaction will not be proportional to the concentration N of the catalyst. Such a proportionality will be found only at small concentrations of the catalyst. With increasing concentration of the catalyst, the rate of reaction will tend to a limiting constant value. This type of dependence of the rate of reaction on the concentration of the catalyst was often observed experimentally in the hydrogenation of unsaturated compounds in the liquid phase.

Evidently, this method is applicable also to more complex cases when the true reaction kinetics at the surface of the disperse particles does not follow a first order.

In the absence or in the presence of very weak convection in the liquid, one must take into account the decrease of the concentration C with increasing distance from the surface. In this case, one can use directly the previously discussed results of Zeldovich's theory which is applicable not only to a reaction on a porous surface but also to all cases when the reaction takes place in the volume, but the reactant is supplied by diffusion from the other phase.

In particular, in a fairly broad range, the rate of reaction should be proportional to the square root of the catalyst concentration. In the case of microheterogeneous processes linked with the diffusion of one reactant from another phase, there can be a great number of different cases where the rate of the process is limited by different stages and is determined by different parameters. Both diffusion in the volume and diffusion from the volume to the surface of the disperse particles can be essential for the rate of the process.

The role of the diffusion to the surface is determined by the ratio of the magnitudes k and β. The role of the diffusion in the volume, in the presence of sufficiently strong convection when formula (II, 82) is applicable, is determined by the ratio of the magnitudes $\frac{D}{\delta}$ S and k'ω. If convection is absent or is only very weak and if Zeldovich's theory is applicable, the role of the diffusion in the volume will be determined by the ratio between δ and the depths of penetration L of the reaction into the volume of the liquid phase. The latter is given by formula (II, 70), which in this case takes the form

$$L = \sqrt{\frac{D}{k'}} = \sqrt{\frac{D(k + \beta)}{N\sigma k\beta}} \qquad\qquad (II, 83)$$

In complete absence of convection, the role of the thickness δ will be played by the thickness of the layer of the liquid phase or the linear dimension of the system.

If $\frac{D}{\delta} S \gg k'\omega$ or $L \gg \delta$, diffusion in the volume is unessential. Then, if $k \gg \beta$, the rate of the process at the surface of the disperse particles will be entirely determined by the diffusion to that surface. In contrast, at $k \ll \beta$, neither diffusion in the volume nor diffusion to the surface will be essential, and the rate will be entirely determined by the true kinetics on the surface.

With weak diffusion and $L \ll \delta$, when diffusion in the volume is essential, both the case $k \gg \beta$, in which diffusion to the surface is essential, and the case $k \ll \beta$, in which the true kinetics at the surface is essential, are possible.

In the presence of strong convection, when formula (II, 82) is applicable, the rate of the process in either case will be entirely determined by diffusion in the volume, i.e., will be simply equal to the rate of dissolution of the reacting gas. Therefore, in the presence of strong convection, it is reasonable to speak only of three characteristic regions of the course of the reaction

(1) the region of diffusion in the volume

$$\left(\frac{D}{\delta} S \ll k'\omega\right)$$

(2) the region of diffusion to the surface

$$\left(\frac{D}{\delta} S \gg k'\omega; \; k \gg \beta\right)$$

(3) the kinetic region

$$\left(\frac{D}{\delta} S \gg k'\omega; \ k \ll \beta \right)$$

Between any two of these regions, there exists a transition region.

One can carry the process from the diffusional into the kinetic region with respect to diffusion in the volume regardless of where the diffusion to the surface lies. Diffusion in the volume and diffusion to the surface depend on different factors. Thus, by changing the concentration of the catalyst, we change the conditions of the diffusion in the volume, and we can carry the process from the kinetic into the diffusional region with respect to diffusion in the volume without in any way changing the ratio of the magnitudes k and β, i.e., without influencing the conditions of the diffusion to the surface. Dispersion of the reacting gas, i.e., increase of the surface of contact between the gaseous and the liquid phase, also influences only the conditions of the diffusion in the volume and, consequently, leads to an increase of the rate of reaction only in the range of diffusion in the volume, where the rate of the reaction is proportional to that surface of contact.

Literature

1. PANETH and HERZFELD, Z. Elektrochem. **37**, 577 (1931).

2. DAMKOHLER, Der Chemie-Ingenieur, III, 2 (1937).

3. PREDVODITELEV and TSUKHANOVA, Zhur. Tekh. Fiz. **10**, 1113 (1940).

4. LEVICH, Zhur. Fiz. Khim. **18**, 335 (1944).

5. FRANK-KAMENETSKII, Zhur. Fiz. Khim. **13**, 756 (1939).

6. FISCHBECK, Z. Elektrochem. **39**, 316 (1933); **40**, 517 (1934).

7. TU, DAVIS and HOTTEL, Ind. Eng. Chem. 26, 749 (1934).

8. ZELDOVICH, Zhur. Fiz. Khim. 13, 163 (1939).

9. VULIS, Issledovanie protsessov goreniya naturalnogs topliva (investigation of combustion processes of natural fuel), ed. Knorre; Moscow (in press).

10. ROITER, Zhur. Fiz. Khim. 14, 1229 (1940).

11. PARKER and HOTTEL, Ind. Eng. Chem. 28, 1334 (1936).

12. Protsess goreniya uglya (the combustion process of coal), ed. Predvoditelev, Moscow 1938.

13. FRANK-KAMENETSKII, Uspekhi Khim. 7, 1277 (1938).

14. FRANK-KAMENETSKII, Zhur. Tekh. Fiz. 10, 1207 (1940).

15. KING, J. Am. Chem. Soc. 57, 828 (1935); 59, 63 (1937); 61, 2290 (1939); KING and HOWARD, Ind. Eng. Chem. 29, 75 (1937).

16. UCHIDA and NAKYAMA, J. Soc. Chem. Ind. Japan, (Suppl.) 36, 635 (1933).

17. BUBEN and FRANK-KAMENETSKII, Zhur. Fiz. Khim. 20, 225 (1946).

18. DRINBERG, Zhur. Fiz. Khim. 6, 871 (1932).

19. GARDNER and KAPPENBERG, Ind. Eng. Chem. 28, No. 4, (1936).

20. HOLDEN and PRIESTLEY, Respiration. Russian translation, p. 237. Moscow, 1937.

21. EVANS, Corrosion of Metals, Russian translation, Moscow-Leningrad, 1941.

22. VULIS, Zhur. Tekh. Fiz. 10, 1959 (1940).

23. MOTT and GURNEY, Electronic processes in ionic crystals. Oxford, Clarendon Press, 1940.

24. PLETENEV AND SOSUNOV, Zhur. Fiz. Khim. 13, 901 (1939).

25. FRANK-KAMENETSKII, Zhur. Fiz. Khim. 13, 1403, (1939).

26. CHUKHANOV and GROZDOVSKII, Zhur. Priklad. Khim. No. 8, p. 1398 (1934); No. 1, p. 73 (1936); Khim. tverd. Topl. No 9-10, p. 902, 986 (1936).

27. FRANK-KAMENETSKII, Zhur. Tekh. Fiz. 9, 1457,
     (1939).

28. FAIRLEY, J. Chem. Soc. 31, 5 (1877).

29. DAVIS, J. Am. Chem. Soc. 52, 3757, 3769, (1930);
     54, 2340 (1932).

30. ELOVICH and ZHABROVA, Zhur. Fiz. Khim. 19, 239,
     (1945); GOLDANSKII and ELOVICH, Zhur. Fiz. Khim.
     20, 1085 (1946); ELOVICH and ZHABROVA, Kinetika
     kataliticheskogo gidrirovaniya triglitseridov
     zhirnykh kislot (Kinetics of the catalytic
     hydrogenation of triglycerides of fatty acids)
     (in press).

31. SEMENOV, Acta physicochimica. 18, 93, (1943).

32. KLIBANOVA and FRANK-KAMENETSKII, Acta Physicochim.
     18, 387 (1943).

33. VULIS and VITMAN, Zhur. Tekh. Fiz. 11, 509, (1941).

34. VULIS, Zhur. Tekh. Fiz. 16, 83 (1946).

35. SEMECHKOVA and FRANK-KAMENETSKII, Zhur. Fiz. Khim.
     14, 231 (1940).

36. DYAKONOV, Thesis, Energy Inst. Acad. Sci. U.S.S.R.,
     1946.

37. NALBANDYAN and SHUBINA, Zhur. Fiz. Khim. 20,
     1249 (1946).

38. PSHEZHETSKII and RUBINSHTEIN, Zhur. Fiz. Khim. 20,
     1127 (1946).

CHAPTER III: THE STEPHAN FLOW

To this point it was assumed that all species taking part in the reaction diffuse independently of each other. Such an assumption is legitimate for diffusion of electrically neutral particles in solution or for diffusion in a gaseous mixture diluted by a large excess of an inert gas. However, cations and anions cannot diffuse independently of each other, and produce an electric current, since diffusion at different rates gives rise to a potential difference. Such phenomena are very important for electrochemistry (theory of diffusion potentials) but will not be considered in detail here.

In diffusion in gaseous mixtures, the components cannot diffuse independently of each other; in fact, diffusion at different rates must give rise to pressure differences and, consequently, produce mass flow.

Its significance was first pointed out by Stephan[1]; it is therefore referred to as the Stephan flow. It is easily seen that if a heterogeneous reaction is accompanied by a change of volume, it must give rise to a mass flow perpendicular to the surface at which the reaction takes place. Closer examination will show that such a flow can arise also in a reaction involving no change of volume, namely when the diffusion coefficients of the original reactants and the products of the reaction are

128

different. This problem has been analyzed most rigorously in the work of Buben[5].

Consider a chemical reaction represented by a stoichiometric equation written with all its terms in the left-hand member

$$\upsilon_1 A_1 + \upsilon_2 A_2 + \cdots + \upsilon_k A_k + \upsilon_{k+1} A_{k+1} + \cdots = 0$$

where A_1, A_2, ... are the chemical symbols of the original reactants; A_k, A_{k+1}, ... are those of the reaction products; and υ_i are the corresponding stoichiometric coefficients. The stoichiometric coefficients of the reactants will be counted positive and those of the products negative. Let q_i be the total flow of species A_i counted positive when directed to the surface, negative if away. Then the stoichiometric equation of the flow becomes:

$$\frac{q_1}{\upsilon_1} = \frac{q_2}{\upsilon_2} = \cdots = \frac{q_i}{\upsilon_i} = \cdots \qquad (III, 1)$$

This expression coincides formally with (II, 39) but is essentially broader, as it includes also the reaction products.

The total flow of substance q_i is composed of the diffusion flow

$$- D_i \frac{dC_i}{dy}$$

and the mass flow vC_i, where v is the velocity of the mass flow directed perpendicularly to the surface, and y is the coordinate perpendicular to the surface.

In the following we shall use partial pressures rather than concentrations, in order for the results to be applicable also to non-isothermal conditions. Because $p = RTC$, the total flow of substance q_i can be represented in the form

III. THE STEPHAN FLOW

$$q_1 = - \frac{D_1}{RT} \frac{dp_1}{dy} + \frac{v}{RT} p_1 \qquad (III, 2)$$

The quantity $\frac{dp_1}{dy}$ is expressed by:

$$\frac{dp_1}{dy} = v \frac{p_1}{D_1} - RT \frac{q_1}{D_1} \qquad (III, 3)$$

Summation over all the components of the mixture gives

$$\Sigma_1 \frac{dp_1}{dy} = \frac{dP}{dy} = v \Sigma_1 \frac{p_1}{D_1} - RT \Sigma_1 \frac{q_1}{D_1} \cdots \qquad (III, 4)$$

The number of substances A_1 will include not only species taking part in the reaction but also an inert gas which might be present in the reaction mixture; it will be referred to by the subscript o. In that case,

$$\Sigma_1 \, p_1$$

will designate the sum of the partial pressures of all species present in the gaseous mixture, i.e., will be equal to the total pressure P, hence

$$\Sigma_1 \frac{dp_1}{dy} = \frac{dP}{dy}$$

Further calculation depends on the hydrodynamic conditions. If the mass flow perpendicular to the surface is opposed by a hydrodynamic resistance, the law of that resistance will supply a relation between v and $\frac{dP}{dy}$. With $\frac{dP}{dy}$ expressed by v, (III, 4) can be solved with respect to v, and further calculation can be carried out as will be shown later. However, the case where there is a hydrodynamic resistance has no real meaning; usually the flow of gas from the space to the surface is not opposed. The total pressure can, therefore, be considered the same everywhere, i.e.,

$$\frac{dP}{dy} = 0 \qquad (III, 5)$$

This is the condition of absence of a hydrodynamic resistance. If it holds, one will obtain from (III, 4)

$$\Sigma_1 \frac{q_1}{D_1} = \frac{v}{RT} \Sigma \frac{p_1}{D_1} \qquad\qquad (III, 6)$$

All the q_1 being bound by the stoichiometric condition (III, 1), they can be expressed by any one q_1.

The process is always limited by the diffusion of one species, that for which $\frac{\beta C}{v}$ is smallest. Let this species be designated by the subscript 1. On the basis of (III, 1) we have

$$q_1 = \frac{q_1}{v_1} v_1 \qquad\qquad (III, 7)$$

This expression is valid also for the inert gas. Inasmuch as the inert gas is neither produced nor consumed, its flow must be zero. Consequently, (III, 1) holds for all species without exception if the stoichiometric coefficient of the inert gas is taken equal to zero

$$v_0 = 0 \qquad\qquad (III, 8)$$

Substituting (III, 7) in (III, 6) we get

$$\frac{q_1}{v_1} \Sigma_1 \frac{v_1}{D_1} = \frac{v}{RT} \Sigma_1 \frac{p_1}{D_1} \qquad\qquad (III, 9)$$

The expression (III, 9) is fundamental in the theory of the Stephan flow since it relates the velocity v of the mass flow with the flow q_1 of the species which limits the process. With this relation known, the diffusion equation can be integrated for the species designated by the subscript 1.

From formula (III, 9) it follows that the

velocity v is zero, i.e., the Stephan flow absent when and only when

$$\Sigma_1 \frac{v_1}{D_1} = 0 \qquad\qquad (III, 10)$$

By convention, the flow is positive when it is directed toward the surface, and the stoichiometric coefficients are positive for the original reactants and are negative for the products. Since the diffusion is limited by one of the original species, flowing towards the surface, the magnitudes q_1 and v_1 are intrinsicially positive. The same applies also to

$$\Sigma_1 \frac{p_1}{D_1}$$

Consequently, v will be positive, i.e., the Stephan flow directed to the surface, for

$$\Sigma_1 \frac{v_1}{D_1} > 0$$

and negative, i.e., the flow directed away from the surface, for

$$\Sigma_1 \frac{v_1}{D_1} < 0$$

With all diffusion coefficients D_1 equal, this reduces to the obvious result that the Stephan flow is directed to the surface for reactions accompanied by a decrease of the volume ($\Sigma v_1 > 0$) and away from the surface for reactions involving an increase of the volume ($\Sigma v_1 < 0$).

Let us now examine the effect of the Stephan flow on the rate of the reaction. In an accurate calculation, the geometric shape of the surface may prove essential. Usually, however, the Stephan flow is localized in a thin

layer adjacent to the surface ("boundary layer"); therefore, the surface curvature can be disregarded for purposes of an approximate calculation and the surface can
be considered plane. This approximation is termed the
"boundary layer" method. In that approximate solution,
the total flow of matter is considered constant, independent of y; the mass flow velocity v should also be
independent of y. On the other hand, the partial pressures p of the different species (and their concentrations
as well) change with the distance y from the surface.
The diffusion coefficients D_i in a polycomponent gas
mixture are functions of the concentrations and may, therefore, also be dependent on y.

Let formula (III, 9) be written in the form

$$\frac{RT}{v_1}\frac{q_1}{v} = \frac{\sum_i \dfrac{p_i}{D_i}}{\sum_i \dfrac{v_i}{D_i}} \qquad\qquad \text{(III, 9a)}$$

The left-hand member of this equation does not depend on
y; therefore, the right-hand member should not be dependent on y. Consequently, the dependence of the partial
pressures p_i on y should be such that

$$F = \frac{\sum_i \dfrac{p_i}{D_i}}{\sum_i \dfrac{v_i}{D_i}} \qquad\qquad \text{(III, 11)}$$

be constant.

It can now be seen that the total flow of the
species q_1 is proportional to the velocity v of the
mass flow

$$q_1 = \frac{v_1}{RT}\,Fv \qquad\qquad \text{(III, 12)}$$

To interpret the physical meaning of the magnitude F, we introduce the mean diffusion coefficient $\bar{D}$, defined by

$$\Sigma \frac{P_1}{D_1} = \frac{P}{\bar{D}} \qquad\qquad (III,\ 13)$$

and the symbol γ defined by

$$\frac{\bar{D}}{v_1} \Sigma_1 \frac{v_1}{D_1} = \gamma \qquad\qquad (III,\ 14)$$

The magnitude γ is a dimensionless number of the order of unity. In the case of equality of all diffusion coefficients,

$$\gamma = \frac{\Sigma v_1}{v_1} = -\frac{\Delta v}{v_1}$$

represents the volume decrease in the reaction per unit volume of the species A_1 which limits the process. For $\gamma > 0$ the Stephan flow is directed to the surface, for $\gamma < 0$, away from it, and for $\gamma = 0$ it is zero. The mean diffusion coefficient $\bar{D}$ can change with the distance y from the surface. But, substituting the above notations in the expression for F, we find

$$F = \frac{P}{v_1 \gamma} \qquad\qquad (III,\ 15)$$

from which it follows that γ (and P as well) does not depend on y. Substituting (III, 15) in (III, 12), we find for the total velocity of diffusion

$$q_1 = \frac{P}{\gamma RT} v \qquad\qquad (III,\ 16)$$

All magnitudes in this expression are independent of y.

Expressing, in (III, 3), q_1 by (III, 16), we find

$$\frac{dp_1}{dy} = -\frac{Pv}{\gamma D_1} + \frac{v}{D_1}\, p_1 = -\frac{v}{D_1}\left(\frac{P}{\gamma} - p_1\right) \qquad (III, 17)$$

The diffusion coefficient D_1 is a function of the concentrations of the components, specifically of p_1, and cannot therefore be considered constant. However, in simpler cases this concentration dependence of the diffusion coefficient is weak enough to be disregarded in a first approximation.

Integration of equation (III, 17) on the assumption of a constant D_1 gives

$$\ln\left(\frac{P}{\gamma} - p_1\right) = \frac{v}{D_1}\, y + C \qquad (III, 18)$$

where C is an integration constant.

Inasmuch as we have used the boundary layer method, the integration should extend only over the boundary layer, the thickness of which can be taken equal to the effective film thickness $\delta = \frac{d}{Nu}$ (I, 28).

Let $p_1{}^{\circ}$ represent the partial pressure of the substance A_1 in space, and $p_1{}'$ at the surface. Then we have

1) at $y = 0$, $p_1 = p_1{}^{\circ}$
2) at $y = \delta$, $p_1 = p_1{}'$

From the first condition, we find the integration constant C:

$$C = \ln\left(\frac{P}{\gamma} - p_1{}^{\circ}\right)$$

and then formula (III, 18) takes the form

$$\ln \frac{\frac{P}{Z} - p_1}{\frac{P}{\gamma} - p_1^{\,o}} = \frac{v}{D_1} y \qquad (III, 19)$$

or

$$\ln \frac{1 - \gamma x_1}{1 - \gamma x_1^{\,o}} = \frac{v}{D_1} y \qquad (III, 20)$$

where $x = \frac{p}{P}$ is the mole fraction.

From the second condition we can find the value of the mass flow velocity v

$$v = \frac{D_1}{\delta} \ln \frac{1 - \gamma x_1}{1 - \gamma x_1^{\,o}} \qquad (III, 21)$$

With the aid of formula (III, 16), we can pass to the rate of the reaction, i.e., to the total flow of the limiting species q_1:

$$q_1 = \frac{P}{\gamma RT} \frac{D_1}{\delta} \ln \frac{1 - \gamma x_1'}{1 - \gamma x_1^{\,o}} \qquad (III, 22)$$

Substituting for the thickness of the reduced film δ its value from (I, 28), we get

$$q_1 = \frac{Nu D_1}{d} \frac{P}{\gamma RT} \ln \frac{1 - \gamma x_1'}{1 - \gamma x_1^{\,o}} \qquad (III, 23)$$

or

$$q_1 = \beta_1 \frac{P}{\gamma RT} \ln \frac{1 - \gamma x_1'}{1 - \gamma x_1^{\,o}} \qquad (III, 24)$$

If the reaction mixture is strongly diluted by an inert gas or an excess of one of the reactants, so that both

γx_1^O and $\gamma x_1^!$ are small as compared with unity, the effect of the Stephan flow on the rate of the reaction will vanish. In fact, expansion of the logarithm in (III, 23) into a series gives

$$q_1 \approx \beta_1 \frac{P}{\gamma RT} \cdot \frac{\gamma}{P} (p_1^O - p_1^!) = \beta_1 \frac{p_1^O - p_1^!}{RT}$$

and, with $p = RTC$,

$$q_1 = \beta_1 (C_1^O - C_1^!)$$

which is identical with the expression used without the Stephan flow taken into account.

However, in the presence of a high concentration of the species, the diffusion of which limits the process, the Stephan flow can alter the reaction rate considerably.*

In an irreversible reaction in the diffusion region, $p_1^! = 0$ and the rate of the reaction, with the Stephan flow taken into account, will be expressed by

$$q_1 = -\beta_1 \frac{P}{\gamma RT} \ln(1 - \gamma x_1^O) \qquad \text{(III, 25)}$$

Particularly significant is the effect of the Stephan flow on processes of condensation of vapors. In this case, there is only one reacting substance and there are no gaseous reaction products; consequently $\Sigma \upsilon = \upsilon_1$. The gaseous mixture consists of two species, the condensing

---

* A different consideration of the Stephan flow was given by Damköhler[2] who introduces the wholly unsubstantiated assumption

$$\Sigma_1 D_1 \frac{dp_1}{dy} = 0$$

His results therefore cannot be accepted. F.-K.

vapor and the inert gas; but, in the case of a binary
mixture, $D_1 = D_O = \overline{D}$, and, consequently, in this case,
$\gamma = 1$. The partial pressure of the vapor at the surface
is equal to the saturation pressure at the temperature of
the surface

$$p' = p_{sat}$$

In this case, we obtain from (III, 21) the well known
formula of Stephan

$$q = \frac{D}{\delta} \frac{P}{RT} \ln \frac{P - p_{sat}}{P - p^o} \qquad (III, 26)$$

for the rate of condensation of vapors in the presence of
an inert gas.

If the amount of inert gas in the mixture tends
to zero, p^o will tend to P, and the rate of condensation expressed by Stephan's formula will tend to infinity.

Actually this means that at low concentrations
of the inert gas the rate of condensation is no longer
governed by the diffusion of the condensing vapor to the
surface but by other stages of the process. Practically,
that stage is usually the removal of the heat evolved in
the condensation; that factor is examined in detail in
Nusselt's theory of film condensation.

In reactions involving formation of gaseous
products, the velocity of the Stephan flow can never become infinite, and the change of the rate of the process
cannot be very great due to the Stephan flow.

The Maxwell-Stephan Method

A very elegant approach to diffusion processes
in gases, differing somewhat in form from the preceding,
was proposed by Maxwell and extensively applied by Stephan[1].

So far, we have been treating separately the
molecular flow D grad C and the mass flow vC. In the
molecular-kinetic analysis of diffusion phenomena in
gases, it is difficult to adhere to such a separation since
diffusion is unavoidably accompanied by mass flow.

The Maxwell-Stephan method deals exclusively
with the total flow of matter q, without separating it
into molecular and mass flow.

In the absence of diffusion, the amount of substance transferred by the mass flow is related with the
linear flow velocity v by

$$q = v_n C$$

The index n refers to the vector component
perpendicular to the surface.

In the presence of diffusion one can introduce
a mean linear velocity $\bar{u}$ of motion of the molecules, related with the total flow of substance, q, in the same
manner as the linear velocity v of the mass flow is related with the amount of substance q transferred by
that flow,

$$\bar{u}_n = \frac{q}{C} \qquad\qquad (III, 27)$$

This mean velocity of molecular motion is the basic magnitude in the Maxwell-Stephan method. It must not be confused with the root mean square velocity

$$\sqrt{\overline{u^2}}$$

of the kinetic gas theory. The root mean square velocity
is a scalar, whereas u is a vector.

At equilibrium, i.e., in the absence of diffusion
and of mass flow, $\bar{u}$ = 0, but

$$\sqrt{\overline{u^2}}$$

is by no means equal to zero.

In the foregoing, we made use of the expression for the total flow of substance

$$q = - D(\text{grad } C)_n + v_n C \qquad (\text{III, 28})$$

We shall write such expressions for two different components of the mixture which we shall designate by the subscripts 1 and 2.

$$q_1 = - D_1 (\text{grad } C_1)_n + v_n C_1$$

$$q_2 = - D_2 (\text{grad } C_2)_n + v_n C_2$$

With the aid of formula (III, 27), we find for the mean velocities of molecular motion

$$\bar{u}_1 = - \frac{D_1}{C_1} \text{grad } C_1 + v$$

$$\bar{u}_2 = - \frac{D_2}{C_2} \text{grad } C_2 + v$$

Subtracting the second from the first equation, we have

$$\bar{u}_1 - \bar{u}_2 = \frac{D_2}{C_2} \text{grad } C_2 - \frac{D_1}{C_1} \text{grad } C_1 \qquad (\text{III, 29})$$

Thus, the difference of the mean velocities of molecular motion for two components of the mixture is independent of the presence of a mass flow and is determined solely by the conditions of the diffusion.

As long as we expressed the flow by the velocity, it was necessary to designate the latter by the subscript indicating the velocity component along the perpendicular to the surface. The same applied to the gradient. Now

that we have equated two expressions for the flow and have
expressed the velocity by the gradient, the result has become independent of the entirely arbitrary direction of
that perpendicular, and the subscript n has become
superfluous.

In the case of a binary mixture, $D_1 = D_2 = D$
and, because of the constancy of the total pressure

$$\text{grad } C_1 = - \text{ grad } C_2$$

Formula (III, 29) then gives

$$\tilde{u}_1 - \bar{u}_2 = - D \left(\frac{\text{grad } C_1}{C_1} + \frac{\text{grad } C_2}{C_2} \right)$$

or

$$\tilde{u}_1 - \bar{u}_2 = - \frac{C_1 + C_2}{C_1 \cdot C_2} D \text{ grad } C_1 \qquad (III, 30)$$

This is the Maxwell-Stephan formula. We shall refer to
(III, 30) as the diffusion law in the Maxwell-Stephan
form, and to (III, 28) as the diffusion law in the Fick
form.

Maxwell obtained the formula (III, 30) directly
from the kinetic gas theory which, of course, can yield
the usual diffusion formulas by reversing the above
derivation.

Formula (III, 30) enables one to obtain in a
very easy and simple manner Stephan's formula for the rate
of condensation in the presence of an inert gas. Let
subscript 1 refer to the condensing vapor, and subscript
2 to the inert gas. As the latter is neither produced
nor consumed anywhere, the mean velocity of its molecular
motion in the stationary state will be zero

$$\bar{u}_2 = 0 \qquad\qquad (III, 31)$$

Consequently, in the case under consideration

$$\bar{u}_1 = -\frac{C_1 + C_2}{C_1 \cdot C_2} \, D \text{ grad } C_1 \qquad \text{(III, 32)}$$

According to (III, 27) we can immediately get the total flow of condensing vapor, i.e., the rate of condensation

$$q_1 = \bar{u}_n C_1 = -\frac{C_1 + C_2}{C_2} \, D(\text{grad } C_1)_n \qquad \text{(III, 33)}$$

In many textbooks and tables (particularly in the American literature), this formula is given as the general expression for the rate of diffusion instead of as Fick's law. Actually formula (III, 33) applies only to the special case of a binary mixture, one component of which is neither produced nor consumed anywhere.

In condensation on a plane surface

$$(\text{grad } C_1)_n = \frac{dC_1}{dy} = -\frac{dC_2}{dy}$$

Passing from concentrations to partial pressures, we get from (III, 33), because $p = RTC$,

$$q_1 = \frac{P}{P_2} \frac{D}{RT} \frac{dp_2}{dy} \qquad \text{(III, 34)}$$

where q_1 and P do not depend on y. Integration gives

$$\ln p_2 = \frac{RT}{P} \frac{q_1}{D} \, y + B \qquad \text{(III, 35)}$$

where B is the integration constant.

Let the y axis be disposed in the same way as in the foregoing chapter; we then find the boundary conditions

$$\text{at} \quad y = 0, \quad p_2 = P - p_1^{\,o}$$

$$\text{at} \quad y = \delta, \quad p_2 = P - p_{sat}$$

From the first condition we determine B

$$B = \ln(P - p_1^{\,o})$$

and from the second condition we find q_1

$$q_1 = \frac{P}{RT}\frac{D}{\delta} \ln \frac{P - p_{sat}}{P - p_o^{\,1}}$$

which is identical with Stephan's formula (III, 26).

Condensation of Vapors in the Presence of Incondensible Gases

In many technical applications one finds condensation of a vapor strongly diluted with incondensible gases. Examples are the condensation recovery of volatile solvents, the condensation of spent steam in the condensers of steam engines, the condensation of water vapor prior to absorption of nitrogen oxides to obtain concentrated nitric acid, the condensation of ammonia from the nitrogen-hydrogen mixture after synthesis, the condensation of sulfuric acid vapors in the concentration and manufacture of oleum.

The most interesting and frequent case is where the task consists in the fullest possible condensation of the vapor, calling for a considerably lower concentration in the outgoing as compared with the inflowing gas. The method of such a calculation was proposed by us[3] and was made more precise by Amelin[4][*].

There is a deep fundamental difference between the condensation of pure vapors and the condensation of

[*] See also: Colburn, A. P., and Hougen, O. A., Ind. Eng. Chem., 26 1178 (1934).

pure vapors in the presence of a large excess of incondensible gases. In the first instance, the rate of the process is determined by the removal of the evolved latent heat of condensation; in the second, by the velocity of diffusion of the vapor to the surface at which the condensation takes place, across a layer of incondensible gas which forms at that surface.

In the first instance, it is reasonable to calculate by the methods of the theory of heat transfer and to use the heat exchange coefficient as the basic magnitude. This procedure loses its meaning in the second instance, although such uncritical extension of a method suitable only for pure vapors, to vapors diluted with incondensible gases, is not uncommon in technical calculations. Actually, in the presence of incondensible gas, the rate of condensation is determined by the diffusion of the vapor to the surface, and the Stephan flow has to be taken into account; its role is more significant the lower the content of incondensible gas in the mixture. At each point, the velocity of condensation is given by Stephan's formula (III, 26). To find the final results of the process, it is necessary to average that formula over the whole surface at which the condensation takes place.

Let us consider the condensation process in a tube of length L and diameter d. Let P denote the total pressure, p_o the partial pressure of the condensing vapor at the entrance of the tube, and p_f at the exit, and let p_{sat} be the saturation pressure at the temperature of the wall, assumed to be uniform over its whole length: Integration of Stephan's formula over the length of the tube gives the result

$$\ln \frac{\ln \dfrac{P - p_o}{P - p_{sat}}}{\ln \dfrac{P - p_f}{P - p_{sat}}} = Z \qquad\qquad (III, 36)$$

where the magnitude Z depends on the dimensions of the
tube and the conditions of the diffusion. The latter are
best described by Margoulis criterion M with the aid of
which Z is expressed simply by

$$Z = \frac{4L}{d} M \qquad\qquad (III, 37)$$

In the practically most important case of turbulent flow,
Margoulis criterion depends only little on the properties
of the gas and the hydrodynamic flow. Consequently, in the
case of turbulent flow, attainment of a stated degree of
condensation requires a definite ratio between length and
diameter.

  Simultaneously, with the condensation of vapor
at the surface, the gas mixture is cooled down as a result
of heat exchange. If this cooling is too rapid, the vapor
can become supersaturated and volume condensation will set
in. Volume condensation is usually undesirable since it
leads to formation of fine droplets of liquid, which are
carried away by the gases in the form of mist. It is much
more difficult to collect such a mist than to condense the
vapor. Particularly harmful is the formation of misty
sulfuric acid in the processes of concentration of sulfuric
acid and manufacture of oleum. To combat it, it is necessary to calculate correctly the thermal conditions of
the process. The main requirement is that the cooling of
the gas be not too fast in comparison with the diffusion
process. The paradoxical conclusion is that all too intense cooling can defeat the results of the condensation
process.

Literature

1.   STEPHAN, Ann. der Physik, 17, 550 (1882); 41, 725 (1890).

2.   DAMKOHLER, Der Chemie-Ingenieur III, Th. 1, 448 ff.(1937).

3.   FRANK-KAMENETSKII, Zhur. Tekh. Fiz. 12, 327 (1942).

4.   AMELIN, Zhur. Tekh. Fiz. 15, 287 (1945).

5.   BUBEN, Sbornik rabot po fizicheskoi Khimii, (Supple-
        ment to Zhur. Fiz. Khim. 1946). p. 148, 154 (1947).

CHAPTER IV: NONISOTHERMAL DIFFUSION

Equations of Heat Conductance and of Diffusion in the Simultaneous Presence of Both Processes

Up to this point we have dealt with heat conductance in a medium in which the concentration is constant and the same at all points, and we have dealt with isothermal diffusion. In the first instance, only heat transfer processes occur in the system, to the exclusion of diffusion; in the second, only diffusion took place, to the exclusion of heat transfer.

In practice, very frequently both a concentration and a temperature gradient are present in the same system, and heat transfer and diffusion occur simultaneously. The process then becomes more complex, and entirely new phenomena, known as thermal diffusion and diffusion thermoeffect, arise.

This is essentially due to the fact that the heat flow depends not only on the temperature gradient but also on the concentration gradient; and the diffusion flow depends not only on the concentration, but also on the temperature gradient.

This necessitates introduction, in the expression for the heat flow, in addition to the $- \lambda \operatorname{grad} T$ term,

of still another term proportional to the concentration
gradient; and in the expression for the diffusion flow,
in addition to the - D grad C term, of another term proportional to the temperature gradient.

The phenomenon of thermal diffusion has recently become important, as it has become the basis of a very
effective method of separation of isotopes proposed by
Clusius.

The Enskog-Chapman Laws

The laws of thermal diffusion and diffusion
thermoeffect have been derived from the kinetic gas theory
by Enskog and Chapman. A detailed exposition of that
theory will be found in the book of Chapman and Cowling[1].

Its final result is the following expression for
the law of diffusion in the Maxwell-Stephan form:

$$\bar{u}_1 - \bar{u}_2 = - \frac{p^2}{p_1 p_2} D\left[\frac{1}{p}\text{grad } p_1 + \frac{k_T}{T}\text{grad } T \right] \qquad (IV, 1)$$

and the corresponding expression for the heat flow

$$q = - \lambda(\text{grad } T)_n + \bar{I}\sigma_n + JPk_T(\bar{u}_1 - \bar{u}_2) \qquad (IV, 2)$$

Here, k_T is the so-called thermal diffusion ratio, a
dimensionless magnitude characteristic of the given gas
pair as its physical constant (sometimes, the magnitude
$D_T = k_T D$, termed the thermal diffusion coefficient, is
introduced instead; k_T is thus defined as the ratio of
the thermal diffusion and the diffusion coefficient);
$\bar{I} = c_\rho T$ is the mean heat content of the mixture; J is
the thermal equivalent of the work, converting the second
term of the right-hand member of (IV, 2), which has the
dimension of mechanical work, into heat units.

Formula (IV, 1) and all expressions for the

diffusional flow formulated in the Maxwell-Stephan form are valid only for a binary mixture.

Formulas (IV, 1) and (IV, 2) show that thermal diffusion and diffusion thermoeffect phenomena are closely related and that their intensity is determined by the value of the same physical constant k_T, there is no special coefficient of diffusion thermoeffect. This is but one particular instance of a general law of nature, the so-called principle of symmetry of kinetic coefficients, or the Onsager principle.

The numerical value of the thermal diffusion ratio k_T depends on the individual properties of the given gas pair. The kinetic gas theory shows that this magnitude is extremely sensitive to the particular mechanism of collision between molecules, specifically, to the law of the repulsive forces acting between the molecules at close ranges in the act of collision.

It is not enough to specify that the gas is considered ideal, i.e., that interaction forces are negligible at mean intermolecular distances. Even with an ideal gas, the magnitude k_T can have widely different values depending on the law of the interaction forces arising between the molecules at distances much closer than the mean distance, such as are involved in the act of collision.

If the molecules are viewed as material points (force centers) which at close range repel each other with a force inversely proportional to a nth power of the distance, one has for $n = 5$, $k_T = 0$, i.e., thermal diffusion is totally absent. Maxwell in his classic studies of kinetic gas theory considered particularly the case $n = 5$, mathematically the simplest. Consequently, there was no room in his theory for thermal diffusion or diffusion thermoeffect. These phenomena for ideal gases were predicted in 1912 - 1915 by Enskog and Chapman who developed this theory following Maxwell, but for any values of n.

For all ordinary gases n > 5. If, in formula
(II, 1), the subscript 1 refers to the heavier gas (or,
in the case of gases of equal molecular weight, to the gas
with larger molecules), k_T will be a positive magnitude.
With the subscripts reversed, k_T will be negative. This
means that heavier (or, at equal molecular weight, larger)
molecules tend as a result of thermal diffusion to con-
centrate in the colder parts of the system. For n < 5,
the signs are reversed. This case appears to be realized
only in strongly ionized gases. The numerical value of
k_T depends very strongly on the composition of the mixture. At a low concentration of one component of the
mixture, k_T is proportional thereto,

$$k_T = bx \qquad\qquad (IV, 3)$$

where x is the mole fraction of the substance present at
a low concentration. The proportionality coefficient b
does not for any known gas pair exceed 0.2 - 0.3.

With increasing concentration, the growth of
k_T with the concentration becomes slower, k_T passes
through a maximum, and then falls to zero (with decreasing
concentration of the second component).

Figure 17, taken from the book of Chapman and
Cowling, represents the dependence of the thermal diffusion
ratio k_T on the composition for a hydrogen-nitrogen
mixture. The experimentally observed maximum values of
k_T did not exceed 0.1.

The value of k_T is greater, the greater the
difference between the molecular weights of the two gases.
Therefore, thermal diffusion is very significant in mixtures of hydrogen with other gases. It is far less important in pairs of gases with close molecular weights.

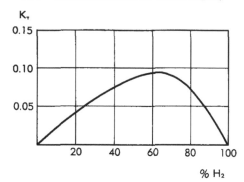

FIGURE 17. DEPENDENCE OF THE THERMAL DIFFUSION
RATIO ON THE COMPOSITION OF A HYDROGEN-NITROGEN
                    MIXTURE
Ordinates are the thermal diffusion ratio,
abscissas are the percent of hydrogen.

The Thermal Diffusion Law in Fick's Form

In order to write down the equations of thermal
diffusion in Fick's form, i.e., to express the flow of
substance in an explicit form, it is necessary to reverse
the transformation through which, in a preceding chapter
[formulas (III, 29) - (III, 30)], we have passed from the
law of diffusion in Fick's form to the law in the Maxwell-
Stephan form, as obtained from the kinetic theory; for
practical application, it must be transformed into the
Fick form.

The total flow of substance is equal to the sum
of the flows of the two components of the mixture

$$\nu (C_1 + C_2) = \bar{u}_1 C_1 + \bar{u}_2 C_2 \qquad (IV, 4)$$

In other words, the linear velocity of the mass
flow represents the weighted mean of the mean velocities
of motion of the molecules

$$\nu = \frac{\bar{u}_1 C_1 + \bar{u}_2 C_2}{C_1 + C_2} \qquad (IV, 5)$$

In order to pass from concentrations to partial

pressures, it is necessary to multiply the numerator and the denominator of the right-hand member of (IV, 5) by RT; this gives

$$\nu = \frac{\bar{u}_1 p_1 + \bar{u}_2 p_2}{P} \qquad\qquad \text{(IV, 6)}$$

where $P = p_1 + p_2$ is the total pressure. With the aid of (IV, 6) we can express $\bar{u}_2$ by $\bar{u}_1$ and ν:

$$\bar{u}_2 = \frac{\nu P - \bar{u}_1 p_1}{p_2} \qquad\qquad \text{(IV, 7)}$$

Substituting this expression for $\bar{u}_2$ in (IV, 1) and assembling all terms containing $\bar{u}_1$ in the left-hand member, we get

$$\bar{u}_1 \left(1 + \frac{p_1}{p_2} \right) = \frac{\nu P}{p_2} - \frac{P^2}{p_1 p_2} D \left[\frac{1}{P} \text{grad } p_1 + \frac{k_T}{T} \text{grad } T \right]$$

Because

$$1 + \frac{p_1}{p_2} = \frac{p_1 + p_2}{p_2} = \frac{P}{p_2}$$

this expression can be rewritten as

$$\bar{u}_1 = \nu - \frac{D}{p_1} \left[\text{grad } p_1 + \frac{p}{T} k_T \text{grad } T \right] \qquad \text{(IV, 8)}$$

We now can express the total flow of the substance designated by the subscript 1 as

$$q_1 = \bar{u}_{1_n} C_1 = \bar{u}_{1_n} \frac{p_1}{RT}$$

$$q_1 = - \frac{D}{RT} \left[(\text{grad } p_1)_n + \frac{p}{T} k_T (\text{grad } T)_n \right] + \frac{\nu_n p_1}{RT} \qquad \text{(IV, 9)}$$

This is the law of diffusion, including thermal diffusion, in the Fick form.

From partial pressures one can pass to concentrations. Because $p = RTC$, and $P = RT\Sigma C$, where ΣC is the sum of concentrations of all the substances, (IV, 9) can be written in the form

$$q_1 = - D(\text{grad } C_1)_n - \left(\frac{k_T \Sigma C + C_1}{T} \right) D(\text{grad } T)_n + \nu_n C_1 \qquad \text{(IV, 10)}$$

since

$$\text{grad } RTC_1 = RT \text{ grad } C_1 + RC_1 \text{grad } T$$

The original Enskog-Chapman formula is applicable, as has been pointed out, only to binary mixtures. But formulas (IV, 9) and (IV, 10) can be extended also to multi-component mixtures; only, in that case, the values of D and k_T will be different for different components and strongly dependent on the composition of the mixture.

Approximate Theory of Nonisothermal Diffusion

For pairs of gases with close molecular weights the numerical value of k_T is small. Consequently, in most practically important cases (with the exception of mixtures containing hydrogen) the term with k_T in (IV, 9) can be disregarded.

The law of nonisothermal diffusion then takes the form

$$q_1 = - \frac{D}{RT}(\text{grad } p_1)_n + \frac{\nu_n p_1}{RT} \qquad \text{(IV, 11)}$$

Thus, if $k_T \ll 1$, the ordinary diffusion equations can be used, without allowance for thermal diffusion, also in the case of nonisothermal diffusion; only, instead of concentrations, one must introduce partial pressures, and one must replace D and ν by $\frac{D}{RT}$ and $\frac{\nu}{RT}$.

This is the most widespread method of calculation of processes of nonisothermal diffusion, and the one most generally used by many authors. It should, however, be borne in mind that this method is legitimate only if k_T is very small, i.e., when the molecular weights of all components differ little.

One can see from formula (IV, 10) that, with larger values of k_T, the error committed through disregarding the thermal diffusion term is of the same order of magnitude as the error of using concentrations instead of partial pressures. Numerically, however, the former error is actually smaller, as the numerical value of k_T is never greater than 0.1.

As was pointed out, if $C_1 \ll \Sigma C$,

$$k_T = bx_1 = b\frac{C_1}{\Sigma C}$$

formula (IV, 10) then becomes

$$q_1 = -D(\text{grad } C_1)_n - \frac{C_1(b+1)}{T}D(\text{grad } T)_n + \nu_n C_1 \quad \text{(IV, 12)}$$

In the calculation of the diffusion of the lighter gas, k_T and, consequently, b, are negative. But the absolute value of b is never in excess of 0.2 - 0.3; consequently, if we disregard the second term in parentheses in (IV, 10), we always commit a greater error than if we disregard the thermal diffusion term in (IV, 9). In other words, use of the gradient of partial pressure in nonisothermal diffusion is always more accurate than the use of the concentration gradient.

Approximate Form of the Law of
Diffusion Thermoeffect

With small k_T we can always use as a first approximation the simple expression of Fourier's law for the heat flow, inasmuch as the additional term which stands in formula (IV, 2) is proportional to k_T.

As a second approximation, we can substitute in (IV, 2) the expression (IV, 1) for $\bar{u}_1 - \bar{u}_2$, and disregard the term containing k_T^2; the expression for the heat flow then takes the form

$$q = - \lambda(\text{grad } T)_n + c\rho T \nu_n - J\, Dk_T \frac{p^2}{p_1 p_2}(\text{grad } P_1)_n \quad (IV, 13)$$

This expression is suitable only for a binary mixture. In the case of a multicomponent mixture, in which only one gas diffuses, one can treat the mixture of all the other gas as an inert gas; p_1 will then mean the partial pressure of the diffusing gas, and p_2 will be $P - p_1$.

In the simultaneous diffusion of several gases, the laws of the thermal diffusion and of diffusion thermoeffect become more complex. Equations of the type of (IV, 1) and (IV, 2) for this case were recently derived by Hellund[2]. Because of the unusual complexity of these expressions, their practical utilization is difficult.

Differential Equations of Simultaneous
Heat Conductance and Diffusion

Using the common method of derivation of the differential equations of heat conductance and diffusion, but replacing the laws of Fourier and of Fick by the formulas (IV, 13) and (IV, 9), we obtain the differential equations of heat conductance and diffusion in an immobile medium in the following form:

$$c\rho \frac{\partial T}{\partial t} = \text{div} \left[\lambda \text{ grad } T + J\, Dk_T \frac{p^2}{p_1 p_2} \text{ grad } p_1 \right] \quad (IV, 14)$$

$$\frac{\partial C_1}{\partial t} = \operatorname{div}\left[\frac{D}{RT}\operatorname{grad}\,p_1 + Dk_T\,\frac{P}{RT^2}\operatorname{grad}\,T\right] \quad (IV,\ 15)$$

With the aid of (IV, 10), the latter equation can be also expressed in concentrations:

$$\frac{\partial C_1}{\partial t} = \operatorname{div}\left[D\operatorname{grad}\,C_1 + \frac{D}{T}(k_T\Sigma C + C_1)\operatorname{grad}\,T\right] \quad (IV,\ 16)$$

If the differences of temperatures and concentrations in the system are small, the temperature and concentration dependence of the physical constants can be disregarded. If the temperature dependence of the diffusion coefficient can be disregarded, it is even more legitimate to consider $\frac{D}{T}$ or $\frac{D}{T^2}$ as temperature independent. In fact, the diffusion coefficient of a gas being proportional to the absolute temperature to the power 1.5 or 2, the magnitudes $\frac{D}{T}$ and $\frac{D}{T^2}$ are even less temperature dependent that D itself.

On that basis, we can represent our differential equations in the form

$$\frac{\partial T}{\partial t} = a\Delta T + J\,\frac{Dk_T P^2}{c\rho}\operatorname{div}\left(\frac{\operatorname{grad}\,p_1}{p_1 p_2}\right) \quad (IV,\ 17)$$

$$\frac{\partial C_1}{\partial t} = \frac{D}{RT}\,\Delta p_1 + \frac{Dk_T P}{RT^2}\,\Delta T \quad (IV,\ 18)$$

At low concentrations of the diffusing species $(p_1 \ll P)$, i.e., for mixtures strongly diluted by an inert gas, p_2 can be considered constant $(p_2 \approx P)$. In that case, the equations of heat conductance and of diffusion can be represented in a symmetrical form. To that end, we consider $\frac{D}{T}$ a constant, and, since for any x

$$\text{grad ln } x = \frac{\text{grad } x}{x}$$

we get

$$\frac{\partial T}{\partial t} = a\Delta T + J \frac{Dk_T P}{c\rho} \Delta\ln p_1 \qquad (IV, 19)$$

$$\frac{\partial C_1}{\partial t} = \frac{D}{RT} \Delta p_1 + \frac{D}{RT} Pk_T\left[\Delta\ln T - \frac{1}{T}(\text{grad } T)^2\right] \qquad (IV, 20)$$

In the presence of convection, these equations should be supplemented by the usual convection terms ν grad T   and ν grad C_1 of $\frac{\nu}{RT}$ grad p_1.

If the temperature or concentration differences are large, it becomes necessary to·allow for the temperature and concentration dependence of the physical constants. This involves complicated computations.

Literature

1.   CHAPMAN and COWLING, Mathematical theory of non-uniform gases. Cambridge 1940.

2.   HELLUND, Phys. Rev. 57, 319, 328 (1940).

CHAPTER V: CHEMICAL HYDRODYNAMICS

If the transfer of matter is related directly to turbulent motion of the liquid or gas, the investigation of the kinetics of chemical processes in the diffusional region can serve as a means for the study of the hydrodynamic characteristics of the turbulent flow: velocity distribution, pulsation eddy, local structure of the turbulence. This chapter of macroscopic kinetics we call chemical hydrodynamics.

Work in this direction is only in its beginnings. What will be discussed in the following, is mainly a program of future investigations, with only a few preliminary results pertaining to the kinetics of dissolution processes. At the present time, we can point out two main problems of chemical hydrodynamics:

1. Investigation of the velocity distribution near the solid surface (in the so-called laminar sublayer) by the study of the diffusional kinetics of chemical processes at the surface of an immobile solid body.

2. Investigation of the local structure of turbulence by the kinetics of processes taking place at the surface of suspended particles.

It might also be of considerable interest to in-
vestigate transfer processes at the free boundary between
a gas and a liquid, or a liquid and a liquid, where the
velocity distribution could be different from that at a
solid surface. To this end, it would be necessary to make
a detailed study of the diffusional kinetics of the solution of gases in liquids under definite hydrodynamic conditions. However, the experimental material available
for that purpose is far from sufficient.

Convective Diffusion in Liquids and Velocity Distribution Near a Solid Surface

If a chemical process takes place in the diffusional region, its rate is determined by that of the transfer
of matter to the surface where the reaction occurs. If
that process takes place under more or less well defined
hydrodynamic conditions, its investigation permits defined hydrodynamic conclusions.

Until recently, all our knowledge of processes
of convective transfer was derived from studies of heat
transfer. Vast experimental material has been accumulated
in this field, and is being commonly made use of in hydrodynamics and in the theory of turbulence. We set ourselves
the task of utilizing for the same purposes, experimental
data of transfer of matter, i.e., diffusional kinetics of
such simple processes as dissolution[1]. Study of convective diffusion phenomena in gases yields little new in
comparison with transfer of heat, as both instances involve values of the Prandtl criterion of the order of unity.
In contrast, diffusion in liquids, with its very low diffusion coefficients, easily involves Prandtl numbers of
the order of several thousands and more. The diffusion
coefficient of substances dissolved in water is of the
order of 10^{-5} cm^2/sec, which, with a kinematic viscosity
of pure water of 10^{-2} cm^2/sec gives a Prandtl criterion
of about 10^3 even in dilute solutions. In highly

concentrated solutions, the kinematic viscosity is markedly higher. At the same time, there is a strong decrease
of the diffusion coefficient which, in a first approximation, is inversely proportional to the viscosity; thus,
the diffusional Prandtl criterion increases roughly proportionally to the square of the viscosity. By increasing
the viscosity of a ferric chloride solution through addition of calcium chloride, we were able[1] to attain a value
of the diffusional Prandtl criterion of about 40,000.

In this way, study of transfer of matter phenomena in a liquid medium, specifically in aqueous solutions,
has permitted investigation of the limiting case of convective diffusion at very high values of the Prandtl criterion. This is of great interest not only from the chemical but also from the hydrodynamic point of view. The
best approach to that problem is the utilization of data
of kinetics of dissolution[1].

In heat transfer phenomena such high values of
the Prandtl criterion could never be attained. The highest values of the Prandtl criterion, observed in the study
of heat transfer in viscous oils, did not exceed a few
hundred. At that, the accuracy of the results was very
poor owing to many basic difficulties, linked with the
strong temperature dependence of the viscosity of liquids.
Heat transfer phenomena must of necessity be studied under
nonisothermal conditions, where the values of the viscosity and, consequently, also of the Prandtl criterion,
are different at different points; this renders the interpretation of experimental results extremely difficult.
In particular, the heat exchange coefficients on heating
and on cooling are different, and it is necessary to resort to extrapolation to an infinitely small temperature
difference[3] or to rather artificial and poorly substantiated ad hoc methods of recalculation of experimental
data[2].

Investigation of processes of convective diffusion

and, in particular, of the kinetics of dissolution, is free
from these difficulties. It turns out to be the best
method of study of processes of transfer of matter at high
values of the Prandtl criterion under strictly isothermal
conditions. The higher the value of the Prandtl criterion
at which the convection process is studied, the deeper
does one penetrate into the structure of the boundary layer
in closest vicinity to the solid surface.

Diffusional Layer

At a distance from the surface, the transfer of
matter by the mass flow is very intense, and the concentration is completely equalized everywhere. All of the
diffusional resistance is concentrated in the immediate
vicinity of the surface, where convective transfer is
negligible as compared with molecular diffusion. This
zone is sometimes called the diffusional boundary layer
or simply the diffusion layer. Its thickness can be considered tantamount with the effective film thickness discussed in Chapter I.

The greater the Prandtl criterion, the smaller
is the intensity of molecular transfer, and the smaller
must be the coefficient of turbulent exchange for the
transfer of matter by mass flow to be negligible in
comparison with molecular transfer. This means that the
thickness of the diffusion layer decreases with increasing
Prandtl criterion.

With very high values of the Prandtl criterion,
even an insignificant turbulence (or, generally speaking,
a normal component of the velocity) will transfer matter
with a very much higher intensity than molecular diffusion.
Only that zone in which convective (specifically, turbulent)
transfer is entirely absent, will play the role of a diffusion zone. By raising the value of the Prandtl criterion, we move the zone in which the diffusion process
is studied increasingly close to the solid surface.

Investigation of the dependence of the transfer of matter process on the Prandtl criterion, will elucidate the variation of the coefficient of turbulent exchange with the distance from the solid surface. The greater the value of the Prandtl criterion, the closer is the approach to that surface and the deeper the penetration into the boundary layer.

The concept of a diffusion layer is purely conventional, corresponding exactly to the concept of effective film introduced in Chapter I. Physically, the diffusion layer is no different from the rest of the flow. It represents the zone in which the diffusion of the substance under consideration can be looked upon as purely molecular. It is therefore natural for the diffusion layer to have different thicknesses for different substances diffusing in the same flow. The smaller the diffusion coefficient, i.e., the greater Prandtl's criterion, the thinner is the diffusion layer. In other words, it will be necessary to approach the surface that much closer in order to be able to disregard convective diffusion in comparison with molecular diffusion.

The real physical structure of the flow cannot, of course, in any way depend on Prandtl's criterion which has different values for the diffusion of different substances in the same flow. The thickness of the diffusion layer stands in no relation to the real physical structure of the flow, and it represents a purely conventional auxiliary magnitude. One could dispense with it, and use only the Nusselt or Margoulis criteria in its place.

Laminar Sublayer

The concept of a laminar sublayer has an entirely different character. This concept corresponds to a definite representation of the physical structure of the flow. Different assumptions can be made about the nature of the fall of the turbulent exchange coefficient on approaching the solid surface:

1. One can assume that the turbulent
    exchange coefficient vanishes only
    at the very surface of the solid and
    that it has a finite value at even
    the smallest distance therefrom.

2. One can assume that the turbulent
    exchange coefficient vanishes at a
    certain finite distance from the
    solid surface, and that convective
    diffusion is entirely absent at small-
    er distances from the surface. That
    is the zone referred to as the laminar
    sublayer.

The assumption of the existence of a laminar sublayer of finite thickness is a hypothesis requiring experimental verification. The first of the two above-
mentioned hypotheses, according to which the turbulent
exchange coefficient vanishes only at the very surface of
the solid, may be just as legitimate. A choice between
the two hypotheses can be made only on the basis of experimental data on convection at very high values of the
Prandtl criterion. If a laminar sublayer does not exist,
the thickness of the diffusion layer should decrease continuously with increasing Prandtl number. If it does exist, the thickness of the diffusion layer should decrease
only as long as it remains greater than the thickness of
the laminar sublayer. At the limit, at very high values
of the Prandtl criterion, the diffusion layer should coincide with the laminar sublayer. At that point, its
thickness should cease to decrease with further increasing
Prandtl number, and be solely governed by the hydrodynamic
characteristic of the flow. In this limiting case, the
concept of a diffusion layer, or of a reduced film, will
be no longer conventional, but will become identical with
the physical laminar sublayer.

Thus, the study of convective diffusion in liquids is linked with the very interesting physical problems of the nature of the decay of turbulence on approaching the solid surface, and of the law of the change of the turbulent exchange coefficient in its immediate vicinity.

Dimensional Analysis

Let us consider, following von Karman[4], the velocity distribution in the vicinity of the solid surface from the point of view of the theory of similitude. Phenomena taking place in close vicinity of the surface should be governed by values of magnitudes pertaining to that region, not to the main flow. Such magnitudes are the kinematic viscosity of the liquid v, its density ρ, and the tangential stress at the surface τ_o.

From these magnitudes one can construct only one magnitude of the dimension of a length, namely,

$$\delta' = \frac{v}{\sqrt{\dfrac{\tau_o}{\rho}}} \qquad (V, 1)$$

termed the thickness of the boundary layer.

At greater distances from the surface, the root mean square pulsation velocity u tends to a constant value u_o. According to Prandtl, this value is equal to the change of the mean velocity over a length equal to the mixing length

$$u_o = 1 \frac{dv}{dy}$$

where y is the distance from the surface, 1 the mixing length and v the mean velocity. In the turbulent flow, the turbulent exchange coefficient plays the role of the kinematic viscosity

$$A = lu$$

and the magnitude

$$\rho A = \rho l u$$

plays the role of dynamic viscosity.

The tangential stress will be expressed by

$$\tau = \rho A \frac{dv}{dy} = \rho u l \frac{dv}{dy}$$

Substituting u for $l \frac{dv}{dy}$, we get

$$\tau = \rho u^2 \qquad\qquad (V, 2)$$

If the curvature of the surface is disregarded, the tangential stress τ is independent of the distance y from the surface. The magnitude τ, as defined by (V, 2) can then be put equal to the tangential stress at the surface

$$\tau_o = \rho u_o^2 \qquad\qquad (V, 3)$$

where u_o is the value of the mean square pulsation velocity at a great distance from the surface, calculated without allowing for the curvature of the surface. Expressing u_o with the aid of (V, 3) by experimentally measurable magnitudes, we get

$$u_o = \sqrt{\frac{\tau_o}{\rho}}$$

This magnitude is the natural yardstick of velocities for the turbulent flow.

It is customary to express the tangential stress through the velocity V of the main flow and the resistance coefficient f, as was done in the Introduction [formula (I, 30)]. We then find for the pulsation velocity

$$u_0 = \sqrt{\frac{\tau_0}{\rho}} = v\sqrt{\frac{f}{2}} \qquad\qquad (V, 4)$$

and for the thickness of the boundary layer

$$\delta' = \frac{v}{u_0} = \frac{v}{v\sqrt{\frac{f}{2}}} \qquad\qquad (V, 5)$$

The latter magnitude is the natural yardstick of length for the velocity distribution in the vicinity of the solid surface.

In particular, the thickness of the laminar sub-layer, if it exists, should be proportional to δ'. In the following description of the velocity distribution at the solid surface, we shall make use of the dimensionless variables

$$v^* = \frac{v}{u} \quad \text{and} \quad y^* = \frac{y}{\delta'}$$

In the limiting range of pure turbulence where the resistance coefficient can be considered constant, the thickness of the laminar sublayer at constant velocity of the flow should be proportional to the kinematic vis-cosity v.

If the laminar sublayer does exist, the diffusion velocity constant, at the limit of large values of Prandtl's criterion, will tend to the value

$$\beta_\infty = \frac{D}{\delta'}$$

where δ^* is the thickness of the laminar sublayer.

We introduce the dimensionless thickness of the

laminar sublayer L, so defined that in accordance with
(V, 5)

$$\delta^* = L \frac{\upsilon}{u_o} = L\delta'$$

(V, 6)

The magnitude L is a universal constant, the limiting
values of the diffusion velocity constant and of Margoulis
criterion at high values of Prandtl's criterion are

$$\beta_\infty = \frac{Du_o}{L\upsilon} = \frac{V}{Pr} \frac{\sqrt{\frac{f}{2}}}{L}$$

(V, 7)

$$M_\infty = \frac{D\sqrt{\frac{f}{2}}}{L\upsilon} = \frac{1}{Pr} \frac{\sqrt{\frac{f}{2}}}{L}$$

In the limiting range of pure turbulence, the resistance
coefficient, and with it the magnitude M Pr, should tend
to a constant value. In a first approximation, the diffusion coefficients of dissolved substances in a liquid
medium are inversely proportional to the kinematic viscosity. Consequently, in this limiting case, the Margoulis
criterion should be inversely proportional to the square
of the kinematic viscosity. The same applies also to the
diffusion velocity constant at constant flow velocity.

General Formulas

        Transfer of heat, matter, and momentum can take
place either through molecular or through turbulent exchange. The intensity of molecular transfer is characterized by the coefficients of diffusion, heat conductivity,
and kinematic viscosity. The intensity of turbulent transfer is characterized by the value of the turbulent exchange
coefficient, defined by formula (I, 15); this coefficient
will henceforth be designated by A. Its value is the
same for the transfer of all the three magnitudes mentioned

above. Thus, as long as we deal with turbulent transfer, there is always complete similitude between diffusion, heat transfer, and resistance. This similitude is infringed only when molecular transfer becomes significant. We shall write the expressions for the diffusion flow, the heat flow, and the tangential stress, separating the terms corresponding to molecular and to turbulent transfer:

$$q = (D + A) \frac{\partial C}{\partial y} \qquad (V, 8)$$

$$q = c\rho(a + A) \frac{\partial T}{\partial y} \qquad (V, 9)$$

$$\tau = \rho(\upsilon + A) \frac{\partial v}{\partial y} \qquad (V, 10)$$

These expressions follow directly from the laws of heat conductance, diffusion, and internal friction, if it is kept in mind that in turbulent transfer the turbulent exchange coefficient A plays the role of diffusion, heat conductivity, and kinematic viscosity, and that

$$\lambda = c\rho a$$

$$\mu = \rho\upsilon$$

Integration gives

$$\Delta C = \int_0^y \frac{q \, dy}{D + A} \qquad (V, 11)$$

$$\Delta T = \frac{1}{c\rho} \int_0^y \frac{q \, dy}{a + A} \qquad (V, 12)$$

$$V = \frac{1}{\rho} \int_0^y \frac{\tau \, dy}{\upsilon + A} \qquad (V, 13)$$

To solve these integrals, it is necessary to know the dependence of the turbulent exchange coefficient A on the

distance from the surface y. The accurate form of this
dependence for small y is not known at the present time,
and we can only make assumptions. With some such relation
between A and y assumed, one can substitute it in
(V, 11) or (V, 12) and carry out the integration; one will
thus find the relation between the heat or diffusion flow
and the temperature or concentration difference, i.e., the
law of forced convection at any value of Pr. Comparison
of the result with experimental data of heat transfer or
diffusion will show to what extent the assumptions made
correspond to reality.

        At large y, the integrands in (V, 11) - (V, 13)
begin to depend on y not only because of the dependence
of A on y, but also as a result of purely geometric
factors. In all cases, except for a plane surface, the
magnitudes q and τ will, on account of these factors,
depend on y. Let us designate the values of these magnitudes at the surface (at y = 0) by the subscript o. We
can then put

$$q = q_0 f(y) \qquad\qquad (V, 14)$$

$$\tau = \tau_0 f(y) \qquad\qquad (V, 15)$$

where $f(y)$ is the same function for both the heat and
diffusion flow and for the tangential stress, its form being determined only by purely geometric factors. Thus, in
the case of a tube, the total heat flow across a cylindrical surface of radius R - y (where R is the radius of
the tube) does not depend on the value of y; the same
applies to the total resistance force referred to that total
surface area. Consequently, for a circular tube

$$f(y) = \frac{R}{R - y} \qquad\qquad (V, 16)$$

Under other geometric conditions, the function $f(y)$ can

have a different form, but it will always tend to unity for small values of y. Substituting (V, 14) and (V, 15) in (V, 13) and introducing the diffusion velocity constant

$$\beta = \frac{q_0}{\Delta C}$$

the heat exchange coefficient

$$\alpha = \frac{q_0}{\Delta T}$$

and the resistance coefficient

$$f = \frac{\tau_0}{\rho V^2/2}$$

we get

$$\frac{1}{\beta} = \int_0^{y_0} \frac{f(y)\ dy}{D + A} \qquad (V, 17)$$

$$\frac{cp}{\alpha} = \int_0^{y_0} \frac{f(y)\ dy}{a + A} \qquad (V, 18)$$

$$\frac{1}{\frac{V}{2}f} = \int_0^{y_0} \frac{f(y)\ dy}{\upsilon + A} \qquad (V, 19)$$

At the upper integration limit we can take the value $y = y_0$ at which the mean values ΔC, ΔT, V, or, generally speaking, the characteristic values of these magnitudes, used in the construction of the similitude criteria, are attained. Introducing the determining dimension d, and replacing y by the dimensionless coordinate

$$\xi = \frac{y}{d} \qquad (V, 20)$$

we get

$$\frac{1}{Nu} = \int_0^{\xi_0} \frac{f(\xi)\, d\xi}{1 + Pr\,\frac{A}{\upsilon}}$$

$$\frac{1}{2\Gamma} = Re \int_0^{\xi_0} \frac{f(\xi)\, d\xi}{1 + \frac{A}{\upsilon}} \qquad\qquad (V,\ 22)$$

The thickness of the diffusion layer (the equivalent film), according to (I, 28) will be expressed by

$$\delta = \frac{d}{Nu} = \frac{D}{\beta} = \int_0^{y_0} \frac{f(y)\, dy}{1 + Pr\,\frac{A}{\upsilon}} \qquad\qquad (V,\ 23)$$

We shall now examine the concrete results obtained on different assumptions about the dependence of the turbulent exchange coefficient A on the distance from the surface y.

Absence of a Laminar Sublayer

Let the turbulent exchange coefficient increase proportionally to the distance from the surface to the power n. The dimensional analysis having shown that in the vicinity of the surface the yardstick of length can be only the magnitude δ', we can put in the most general case

$$A = \kappa\upsilon \left(\frac{y}{\delta\Gamma}\right)^n \qquad\qquad (V,\ 24)$$

where κ is a constant. Substituting in (V, 23) we get

$$\delta = \int_0^{y_0} \frac{f(y)\, dy}{1 + Pr\kappa\left(\frac{y}{\delta\Gamma}\right)^n} \qquad\qquad (V,\ 25)$$

We introduce a new variable x, defined by the condition

$$x^n = \frac{\kappa \, Pr \, y^n}{\delta'^n} \qquad (V, \, 26)$$

Formula (V, 25) then goes over into

$$\delta = \frac{\delta'}{\sqrt[n]{\kappa \, Pr}} \int_0^{x_0} \frac{f'(x) \, dx}{1 + x^n} \qquad (V, \, 27)$$

where $f'(x)$ is some new function.

The upper limit of the definite integral in (V, 27) has the value

$$x_0 = \frac{\sqrt[n]{\kappa \, Pr} \, y_0}{\delta'} \qquad (V, \, 28)$$

Owing to the thickness of the boundary layer being very small in comparison with the dimensions of the flow as a whole, $\delta' \ll y_0$, the value of x_0 will be very large, and we therefore can take infinity as the upper limit of the integral. Thus, as a first approximation

$$\delta \approx \frac{\delta'}{\sqrt[n]{\kappa \, Pr}} \int_0^\infty \frac{f'(x) \, dx}{1 + x^n} \qquad (V, \, 29)$$

The definite integral and κ being universal constants, we have, omitting constant factors,

$$\delta \sim \frac{\delta'}{\sqrt[n]{Pr}} \qquad (V, \, 30)$$

If the thickness δ' of the boundary layer cannot be considered infinitely small, it is necessary to use expression (V, 5) and then the value of x_0 is found to be

$$x_0 = \sqrt[n]{\kappa \, Pr} \; \frac{y_0}{\upsilon} \, V \, (\frac{f}{2})^{\frac{1}{2}} \qquad\qquad (V, 31)$$

or, with y replaced by the dimensionless coordinate ξ, according to (V, 20)

$$x_0 = \sqrt[n]{\kappa} \, \xi_0 \, \sqrt[n]{Pr} \, (\frac{f}{2})^{\frac{1}{2}} \, \frac{Vd}{\upsilon} \qquad\qquad (V, 32)$$

Fusing the constants κ and ξ_0 into one constant

$$a = \sqrt[n]{\kappa} \, \xi^0$$

and considering that $\frac{Vd}{\upsilon}$ is the Reynolds criterion, we get

$$x_0 = a \, \sqrt[n]{Pr} \, (\frac{f}{2})^{\frac{1}{2}} \, Re \qquad\qquad (V, 33)$$

For the thickness of the diffusion layer, formula (V, 29) gives

$$\delta = \frac{\delta'}{\sqrt[n]{\kappa \, Pr}} F(x_0) \qquad\qquad (V, 34)$$

where the function F is of the form

$$F(x) = \int_0^{x_0} \frac{f'(x) \, dx}{1 + x^n} \qquad\qquad (V, 35)$$

Theory of Landau and Levich

The problem of the law of convection on the assumption that the turbulent exchange coefficient vanishes only at the very surface and that a laminar sublayer of finite thickness is absent, was recently examined theoretically by Landau[5] and by Levich[6]. Landau objects to the very term "laminar sublayer", as the existence of a zone where turbulence is wholly absent is impossible in principle.

He refers to the region of the flow immediately adjacent to the surface, as the viscous sublayer. According to Landau, the motion is turbulent even here. The similarity with a laminar flow consists only in the distribution of the mean velocity obeying the same law as the distribution of the true velocity in a laminar flow under the same conditions.

This law is of the form

$$v_x = \frac{\tau_0}{\mu}\, y = \frac{u_0}{\delta'}\, y \qquad\qquad (V, 36)$$

where v_x is the mean velocity, μ the viscosity, y the distance from the surface, τ_0 the tangential stress at the surface; the magnitudes u_0 and δ' are defined by formulas (V, 4) and (V, 5).

According to Landau, the longitudinal component u_x of the pulsation velocity in the viscous sublayer is of the same order of magnitude as the component of the mean velocity, and is proportional to y

$$u_x \cong \frac{\tau_0}{\mu}\, y + \cdots \qquad\qquad (V, 37)$$

The right-hand member of (V, 37) is the first term of a development in a series by powers of y. For small y, the higher terms can be disregarded. The transverse component of the pulsation velocity u_y Landau finds from the continuity equation

$$\frac{\partial u_y}{\partial y} = \frac{\partial u_x}{\partial x}$$

According to (V, 37)

$$\frac{\partial u_x}{\partial x}$$

is proportional to y; hence u_x is proportional to y^2.

This reasoning is valid not only for the transverse
component of the pulsation velocity but, in general, for
the transverse components of the velocity if it exists.
Levich[6], using the equations of the laminar boundary layer,
arrives at the following expression for the transverse
components of the velocity in the vicinity of the surface

$$u_y \simeq \frac{\upsilon}{\delta_1^3} y^2 \qquad\qquad (V, 38)$$

The turbulent exchange coefficient A, according to (I, 15)
is equal to the product of the transverse component of the
velocity u_y and the mixing length 1.

     Landau and Levich assume that in the case of a
laminar boundary layer the mixing length is proportional
to the distance from the surface

$$1 \sim y \qquad\qquad (V, 39)$$

In the case of a turbulent boundary layer, at short distances from the surface (in the viscous sublayer) the
mixing length, according to Landau and Levich, should increase proportionally to the square of the distance from
the surface

$$1 \sim y^2 \qquad\qquad (V, 40)$$

Inasmuch as the transverse component of the velocity u_y
is proportional to y^2, one must assume in formula (V, 30),
for a laminar boundary layer

$$n = 3 \qquad\qquad (V, 41)$$

and for a turbulent boundary layer

$$n = 4 \qquad\qquad (V, 42)$$

Omitting dimensionless factors of the order of unity, we get the results of the theory of Landau and Levich for a laminar flow

$$\delta \cong \frac{\delta'}{\sqrt[3]{Pr}}$$ (V, 43)

and for a turbulent boundary layer

$$\delta \cong \frac{\delta'}{\sqrt[4]{Pr}}$$ (V, 44)

Laminar Sublayer Without Transition Zone

A diametrically opposite model of the phenomenon was proposed in the classic work of Prandtl[7].

In the simplest, roughest approximation, Prandtl split the flow into two regions: a laminar sublayer in which turbulence is entirely absent, and the core of the flow where, on the contrary, molecular transfer can be disregarded against the turbulent exchange. Correspondingly, we put: in the laminar sublayer, $A = 0$, and in the core of the flow, D, a, and v, negligibly small as compared with A.

With the thickness of the laminar sublayer designated by $\delta*$, the integration interval in (V, 17) - (V, 19) or (V, 23) is split into two parts: at $y < \delta*$, the second term in the denominator of the integrand can be disregarded, and at $y > \delta*$, the first term. Formula (V, 23) can now be written as

$$\delta = \int_{0}^{\delta*} f(y) \, dy + \frac{v}{Pr} \int_{\delta*}^{y_0} \frac{f(y) \, dy}{A}$$ (V, 45)

But, since $\delta*$ is small compared with the radius of curvature of the surface, $f(y)$ in the first integral can be taken equal to unity, and the integral is simply equal to

$\delta*$. Computation of the second integral requires the knowledge of the dependence of A on y; we can circumvent
this difficulty by expressing the second integral through
the resistance coefficient. To this end we transform
(V, 19) in analogy to (V, 45), which gives

$$\frac{1}{\frac{V}{2} f} = \frac{\delta*}{\upsilon} + \int_{\delta*}^{y_0} \frac{f(y)\ dy}{A} \qquad (V, 46)$$

Hence, the integral which interests us is

$$\int_{\delta*}^{y_0} \frac{f(y)\ dy}{A} = \frac{1}{\frac{V}{2} f} - \frac{\delta*}{\upsilon} \qquad (V, 47)$$

and, substituting in (V, 45), we have

$$\delta = \delta* + \frac{\upsilon}{Pr}\left(\frac{1}{\frac{V}{2} f} - \frac{\delta*}{\upsilon}\right) \qquad (V, 48)$$

For the diffusion velocity constant we get

$$\beta = \frac{D}{\delta} = \frac{1}{\frac{2}{Vf} + \frac{\delta*}{\upsilon}(Pr - 1)} \qquad (V, 49)$$

and for the Margulis criterion

$$M = \frac{\beta}{V} = \frac{1}{\frac{2}{f} + \frac{\delta*V}{\upsilon}(Pr - 1)} \qquad (V, 50)$$

For Pr = 1, we have $M = \frac{f}{2}$, as it should be according to
Reynold's analogy [formula (I, 36)]. From (I, 48) it is
seen immediately that for Pr = 1, the thickness of the
diffusion layer has nothing to do with the thickness of
the laminar sublayer and does not depend at all on the
value of $\delta*$. On the other hand, at Pr $\gg$ 1, the thickness of the diffusion layer tends to the thickness of the

laminar sublayer.

From dimensional considerations it follows that the thickness of the laminar sublayer should be proportional to the full thickness of the boundary layer, expressed by formula (V, 5). Designating the ratio of these two magnitudes by the term of dimensionless thickness of the laminar sublayer L, and considering it a universal constant, we obtain by substituting the value of $\delta*$ (V, 6) in (V, 5)

$$\delta = \frac{1}{Pr} \frac{2\upsilon}{fV} \left[1 + L\sqrt{\frac{f}{2}} \,(Pr - 1) \right] \qquad (V, 51)$$

Formula (V, 50) then takes the form

$$\frac{1}{M} = \frac{2}{f} + L\sqrt{\frac{2}{f}} \,(Pr - 1) \qquad (V, 52)$$

Prandtl himself prefers a somewhat different form of his formula. He introduces the value of the velocity of flow at the boundary between the laminar sublayer and the core of the flow, i.e., at $y = \delta*$, which he designates by v'. Inasmuch as within the limits of the laminar sublayer the velocity distribution should be linear, the velocity gradient in the laminar sublayer will be expressed by

$$\frac{\tau_0}{\mu} = \frac{v'}{\delta*} \qquad (V, 53)$$

Expressing τ_0 by the resistance coefficient, we get

$$\delta* = \frac{v'\upsilon}{\dfrac{f}{2} V^2} \qquad (V, 54)$$

Substituting in (V, 48) and (V, 50) we get

$$\delta = \frac{1}{Pr} \frac{2}{Vf} \left[1 + \frac{v'}{V}\,(Pr - 1) \right] \qquad (V, 55)$$

$$\frac{1}{M} = \frac{2}{f} \left[1 + \frac{v'}{V} (Pr - 1) \right] \qquad (V, 56)$$

Empirical Formulas

Prandtl's formula has served as a basis for a number of empirical formulas tending to describe the experimental material obtained in the study of heat transfer in viscous liquids. A detailed review of all this material, with a thorough analysis of the methods of recalculation of the experimental data, can be found in the book of Hofmann[2].

Let us then designate the ratio of Prandtl's velocity at the boundary of the laminar sublayer and the mean velocity of the flow by φ

$$\frac{v'}{V} = \varphi \qquad (V, 57)$$

Prandtl's formula can then be written in one of the following forms

$$\delta = \frac{1}{Pr} \frac{2}{V\!f} [1 + \varphi(Pr - 1)] \qquad (V, 58)$$

$$\frac{1}{M} = \frac{2}{f} [1 + \varphi(Pr - 1)] \qquad (V, 59)$$

$$Nu = \frac{\frac{f}{2} Re\ Pr}{1 + \varphi(Pr - 1)} \qquad (V, 60)$$

The majority of the empirical formulas follows the type of Prandtl's formula, with various arbitrary assumptions on the dependence of the magnitude φ on the properties of the flow. As a rough approximation, φ can be considered constant. Ten Bosch[8], from experiments with water, found for φ the value of 0.35.

If the laminar sublayer is considered as an
accurate, and not merely an approximate concept, and its
dimensionless thickness L, a constant magnitude, φ
should be proportional to the square root of the resistance coefficient. This can be seen, for example, from
(V, 51) and (V, 58):

$$\varphi = L\sqrt{\frac{f}{2}} \qquad\qquad (V,\ 61)$$

On this basis, Prandtl derives the dependence of φ on
the Reynolds criterion from Blasius' resistance law
(I, 48) according to which the resistance coefficient is
inversely proportional to the fourth-power root of the
Reynolds criterion. It then follows from (V, 61) that φ
should be inversely proportional to the eighth-power root
of the Reynolds criterion.

Prandtl chooses the proportionality factor so as
best to fit the experimental data, and, as a result, he
obtains for the internal problem (straight-circular tube)
the formula

$$\varphi = 1.74\ \text{Re}^{-\frac{1}{8}} \qquad\qquad (V,\ 62)$$

Combining (V, 62) with (V, 61) and with Blasius' law
(I, 48) according to which

$$\sqrt{\frac{f}{2}} = 0.395\ \text{Re}^{-\frac{1}{8}}$$

we find for the dimensionless thickness of the laminar
sublayer

$$L = 4.4 \qquad\qquad (V,\ 63)$$

If one attributes a real physical existence to
the laminar sublayer, and considers it as a region of
complete absence of turbulence, the magnitude φ should be

determined entirely by the velocity distribution in the
flow, and, consequently, be a function only of the Reynolds
criterion. Several investigators have attempted to choose
the dependence of φ on the Reynolds criterion so that
Prandtl's formula will fit the experimental data satisfactorily. However, in order to attain a good agreement with
the experiment, one must consider φ as a function not
only of the Reynolds criterion but also of the Prandtl
criterion.

Ten Bosch[8] on the basis of a treatment of experimental results obtained with different substances, has proposed for φ the formula

$$\varphi = B\ Re^{-0.11}\ Pr^{-0.185}$$

where $B = 1.4$ on heating and $B = 1.12$ on cooling.
Kupryanov[9] believes that best agreement with the experimental data can be obtained with φ considered as a function
of the Prandtl criterion only:

$$\varphi = 0.44\ Pr^{-\frac{1}{8}}$$

Hofmann[2] proposes the formula

$$\varphi = 1.5\ Re^{-\frac{1}{8}}\ Pr^{-\frac{1}{6}}$$

It should be borne in mind that experimental data of heat
transfer at high values of the Prandtl criterion have been
obtained by measurements of the heat transfer in viscous
oils. The viscosity of such liquids is very strongly
temperature dependent. Therefore, such experiments are
always strongly complicated by the effect of the temperature dependence of the physical constants on the law of
convection. This manifests itself immediately in the experiment by the difference of the heat exchange coefficients
on heating and on cooling.

Therefore, the experimental material available in the literature does not permit a sufficiently accurate establishment of the law of convection at high values of the Prandtl criterion. More accurate results are obtained by means of studies of turbulent diffusion processes under strictly isothermal conditions. Practically, such results can be obtained, for example, in the study of dissolution processes taking place in the diffusional range.

A further refinement of the above rough calculations can be attained by abandoning the coarse model of Prandtl, and representing the flow velocity v as a continuous function of the distance from the surface y, which amounts to linking the laminar sublayer with the core of the flow. Calculations of this kind will be discussed further below. The fact that φ as a function solely of Re fails to render satisfactorily the experimental results, is proof of the inadequacy of the original model of Prandtl, and the necessity of linking the laminar sublayer and the main stream. However, the data available in the literature do not warrant any conclusions about the physical existence of the laminar sublayer, i.e., provide no clue to whether the turbulent exchange coefficient becomes zero at some finite distance from the surface. To answer this question, there is need for experimental data at very high values of the Prandtl criterion.

An extrapolation of empirical formulas obtained at moderate values of this criterion, to very high values, cannot, of course, be considered permissible.

Velocity Distribution in the Turbulent Flow

Any theory pretending to a complete description of the structure and properties of the flow in the vicinity of a solid surface, must be in agreement with the experimental data of velocity distribution.

Extensive and very careful measurements of this

kind for turbulent motion in tubes have been done by
Nikuradze. The results of his measurements are expressed
by the formula

$$\frac{v(y)}{v\sqrt{\frac{f}{2}}} = 5.5 + 5.75 \; \lg \left(y \; \frac{V}{\upsilon} \sqrt{\frac{f}{2}} \; \right) \qquad (V, \; 64)$$

Let us compare this velocity distribution with that following from Prandtl's model.

Integrating equation (V, 10) not to the mean
value of the velocity V but to an arbitrary value v(y),
we get

$$v(y) = \frac{\tau_0}{\rho} \int_0^y \frac{f(y) \; dy}{\upsilon + A} \qquad (V, \; 65)$$

Using Prandtl's model, we split the integrand into two
zones, a laminar sublayer and the core of the flow, and
we get

$$v(y) = \frac{\tau_0}{\rho} \left[\frac{\delta*}{\upsilon} + \int_{\delta*}^y \frac{f(y) \; dy}{A} \right] \qquad (V, \; 66)$$

As a first approximation, disregarding the curvature of the
surface, we put f(y) = 1. For the core of the flow, it is
known from hydrodynamics that the trubulent exchange coefficient is proportional to the distance from the surface.
It is expressed by formula (I, 15).

$$A = lu$$

where l is the mixing length, and u the mean pulsation
velocity.

The mixing length l is proportional to the distance from the surface y

$$1 = \kappa y \qquad \qquad (V, 67)$$

where the proportionality factor κ is a universal constant. The mean pulsation velocity u in the core of the flow can be considered equal to u_o

$$u = u_o = V\sqrt{\tfrac{f}{2}} \qquad \qquad (V, 68)$$

Substituting in (V, 66) and integrating, we find

$$v(y) = \frac{\tau_o}{\mu}\,\delta* + \frac{\tau_o}{\rho\kappa\, V\sqrt{\tfrac{f}{2}}}\ln\frac{y}{\delta*} \qquad \qquad (V, 69)$$

or, introducing the dimensionless thickness of the laminar sublayer L with the aid of (V, 6)

$$v(y) = V\sqrt{\tfrac{f}{2}}\left[L + \frac{1}{\kappa}\ln\left(\frac{yV\sqrt{\tfrac{f}{2}}}{L\upsilon}\right)\right] \qquad (V, 70)$$

For a comparison with (V, 64) we write (V, 70) in the form

$$\frac{v(y)}{V\sqrt{\tfrac{f}{2}}} = L - \frac{1}{\kappa}\ln L + \frac{1}{\kappa}\ln\left(y\,\frac{V}{\upsilon}\sqrt{\tfrac{f}{2}}\right) \qquad (V, 71)$$

Thus, the numerical coefficients of the formula of Nikuradze should be related with the magnitudes involved in Prandtl's theory of the laminar sublayer by

$$\frac{1}{\kappa} = \frac{5.75}{2.3} = 2.5 \qquad \qquad (V, 72)$$

$$1 - 5.75\log L = 5.5 \qquad \qquad (V, 73)$$

The latter equation gives for the dimensionless thickness of the laminar sublayer the value

$$L = 11.7 \qquad (V, 74)$$

The sharp discrepancy between this value and the value
(V, 63) is proof of the inadequacy of Prandtl's model.

Laminar Sublayer With Transition

The basic shortcoming of Prandtl's theory is that
it fails to rest on a continuous distribution of the normal
component of the velocity, but assumes that at the boundary
of the laminar sublayer the pulsation velocity changes discontinuously from zero to a finite value. Such a discontinuous variation is physically impossible. To construct
a physically satisfactory theory, one must introduce a
continuous link between the laminar sublayer and the core
of the flow.

Karman[4] was the first to propose an improvement
on Prandtl's theory, by splitting the flow not into two
but into three zones: a laminar sublayer, the core of the
flow, and an intermediate linking zone. Further similar
calculations, involving different arbitrary assumptions on
the velocity distribution in the linking zone, were made
by Hofmann[2] and by Mattioli[11]. The approach used by these
authors complicates Prandtl's theory considerably, but does
not eliminate its basic shortcoming - the law of distribution of pulsation velocities used in these calculations
consists of three separate segments, but still is not
continuous.

We have considered the problem of the calculation
of convective diffusion processes on the assumption of a
continuous transition from the laminar sublayer to the core
of the flow. To this end, we split the flow, following
Karman, into three zones: a laminar sublayer of thickness
δ*, where the turbulent exchange coefficient is strictly
zero, the core of the flow where the velocity distribution
follows the logarithmic law of Nikuradze (V, 64), and an
intermediate transition zone. On the latter, however, we

impose the condition that it should pass smoothly and continuously into the laminar sublayer on one side and the
core of the flow on the other side. Not only the mean
velocity of flow v, but also its derivative with respect
to the distance from the surface should have no discontinuity.

In the laminar sublayer the velocity distribution
should satisfy the law

$$\frac{dv}{dy} = \frac{\tau_0}{\mu} \qquad (V, 75)$$

In the core of the flow, it should tend to the logarithmic
law of Nikuradze, and, consequently, $\frac{dv}{dy}$ should become
inversely proportional to y.

Let us place the origin of the coordinates at the
outer boundary of the laminar sublayer, and let us set
ourselves the task of describing the velocity distribution
in the intermediate zone and in the core of the flow with
the aid of one single analytical expression. The ex-
pression for $\frac{dv}{dy}$ should at y $\longrightarrow$ 0 give the value (V, 75),
and at larger values of y it should become inversely pro-
portional to y. In order to meet these two conditions,
we represent $\frac{dv}{dy}$ in the form

$$\frac{dv}{dy} = \frac{\tau_0}{\mu} \left[1 - f\left(\frac{\gamma}{y}\right) \right] \qquad (V, 76)$$

where f is an unknown function which we shall call the
"transition function". The constant γ can be termed the
width of the transition zone.

For the sake of convenience in the following calculations we shall introduce the auxiliary variable

$$\eta = \frac{\gamma}{y} \qquad (V, 77)$$

For $\frac{dv}{dy}$ to be continuous, the velocity distribution must satisfy the transition conditions: at the

boundary of the laminar sublayer, with y tending to infinity, $f(\eta)$ should vanish; in the core of the flow,
with η tending to zero, $f(\eta)$ should tend to the limit
$1 - \eta$. Clearly, the simplest transition function satisfying these requirements is

$$f(\eta) = e^{-\eta} \qquad\qquad (V, 78)$$

The expression (V, 78) for the transition function can be
considered a very convenient interpolation formula for the
distribution of velocities in the vicinity of the surface.

But for the most interesting case of large values
of the Reynolds criterion, the general form of the law of
convective diffusion can be obtained accurately, except for
an unknown function of the Prandtl criterion, for any form
of the function $f(\eta)$, on the sole condition that it should
satisfy the transition conditions formulated above.

According to (V, 10) the expression for the turbulent exchange coefficient corresponding to the velocity
distribution (V, 76) is

$$A = \upsilon \, \frac{f(\eta)}{1 - f(\eta)} \qquad\qquad (V, 79)$$

In the core of the flow, $f(\eta)$ becomes $1 - \eta$, with
$\eta \ll 1$, and, consequently,

$$A = \frac{\upsilon}{\eta} = \frac{\upsilon}{\gamma} \, y \qquad\qquad (V, 80)$$

But, according to (V, 67) and (V, 68), in the core of the
flow

$$A = \kappa y u_{o} \qquad\qquad (V, 81)$$

Combining (V, 80) with (V, 81), we conclude that

$$\gamma = \frac{\upsilon}{\kappa u_{o}} \qquad\qquad (V, 82)$$

Thus, the width of the linking zone cannot be chosen arbitrarily, but is bound with the fundamental characteristics
of the turbulent flow. Taking for the natural yardstick of
length, as was done throughout the foregoing discussion,
the magnitude

$$\frac{v}{u_o}$$

we find for the dimensionless width of the linking zone the
magnitude $\frac{1}{\kappa}$. Here κ is the proportionality factor between the mixing length and the distance from the surface,
a fundamental universal constant characteristic of the
properties of turbulence.

Substituting in (V, 23) the expression (V, 79)
for the turbulent exchange coefficient, and bearing in mind
that the origin of the coordinate η now is not at the
surface but at the outer boundary of the laminar sublayer,
we get for the thickness of the diffusion layer

$$\delta = \delta* + \gamma \int_{\eta_o}^{\infty} \frac{1 - f(\eta)}{1 + f(\eta)(Pr - 1)} \cdot \frac{d\eta}{\eta^2} \qquad (V, 83)$$

where η_o is the value of y at the point where the characteristic value of the velocity V is reached (e.g., at
the center of the tube).

Bearing in mind that the corresponding value of
y is equal to half the determining dimension, we get from
(V, 82)

$$\eta_o = \frac{2}{\kappa \sqrt{\frac{f}{2}} \, Re} \qquad (V, 84)$$

Thus, the limiting case of high values of the
Reynolds criterion corresponds to low values of η_o.

In order to obtain results for this most

interesting limiting case, we shall represent the integrand in (V, 83) in the form of a sum of two terms so that,
with η_0 tending to zero, one of them will give an integral expressed by known functions (integral logarithm),
and the other, an integral tending to a finite limit. This
can be attained by introducing an auxiliary function
$\varphi(\eta, Pr)$ so that the integrand in (V, 83) can be represented in the form

$$\frac{1}{Pr}\left[\frac{e^{-\eta}}{\eta} + \varphi(\eta, Pr) \right]$$

Integration of (V, 83) then gives

$$\delta = \delta* + \frac{\gamma}{Pr}\left[- E_1(-\eta_0) + \int_{\eta_0}^{\infty} \varphi(\eta, Pr)\, d\eta \right] \qquad (V, 85)$$

where E_1 is the integral logarithm.

Expressing $\delta*$ and γ with the aid of (V, 6)
and (V, 82) we get for the Margoulis criterion

$$\frac{\sqrt{\frac{f}{2}}}{M\,Pr} = L + \frac{1}{\kappa Pr}\left[- E_1(-\eta_0) + \int_{\eta_0}^{\infty} \varphi(\eta, Pr)\, d\eta \right] \qquad (V, 86)$$

At the limit of large values of the Reynolds criterion,
i.e., of small η_0, the integral in the right-hand member
tends to a constant limiting value which we shall designate
by $F(Pr)$:

$$\int_{0}^{\infty} \varphi(\eta, Pr)\, d\eta = F(Pr.) \qquad (V, 87)$$

Then, with the use of the known series for integral logarithms,
and expressing η_0 with the aid of (V, 84), we get for this
limiting case

$$\frac{\sqrt{\frac{f}{2}}}{M \, Pr} = L + \frac{1}{\kappa \, Pr}\left[F(Pr) - C + \ln 2\kappa + \ln\left(\sqrt{\frac{f}{2}}\, Re\right)\right] \quad (V, \, 88)$$

where $C = 0.577$ is Euler's constant.

Integration of (V, 76) gives for the velocity distribution at great distances from the surface, in analogy to the derivation (V, 88)

$$v^* = \frac{v}{U_0} = L + \frac{1}{\kappa}(G - C + \ln\kappa) + \frac{1}{\kappa}\ln\left(\frac{yv\sqrt{\frac{f}{2}}}{v}\right) \quad (V, \, 89)$$

and for the resistance coefficient

$$\frac{1}{\sqrt{\frac{f}{2}}} = L + \frac{1}{\kappa}\left[G - C + \ln 2\kappa + \ln\left(Re\sqrt{\frac{f}{2}}\right)\right] \quad (V, \, 90)$$

where $G = F(1)$.

Combining (V, 88) with (V, 90) we get, finally,

$$\frac{1}{M} = \frac{2}{f} + L\sqrt{\frac{2}{f}}\left[Pr - 1 + \frac{1}{\kappa L}(F(Pr) - G)\right] \quad (V, \, 91)$$

All theories which split the flow into three regions and introduce a zone intermediate between the laminar sublayer and the core of the flow, can be reduced to this form. If, however, the velocity distribution in the intermediate zone does not satisfy the linking conditions, the dimensionless width of that zone becomes entirely arbitrary and is not obligatorily equal to $\frac{1}{\kappa}$. Von Karman's formula can be viewed as the special case (V, 91) with

$$L = \frac{1}{\kappa} = 5$$

$$\frac{1}{\kappa L} [F(Pr) - G] = \ln[1 + \frac{5}{6} (Pr - 1)] \qquad (V, 92)$$

However, the form of transition function adopted by von Karman,

$$f(\eta) = 1 - 2\eta$$

does not satisfy the above formulated transition conditions and therefore cannot be adjudged correct in principle. The same applies to other more complex forms of the linking. function encountered in the literature.

As a very simple form of the $f(\eta)$ function, satisfying the linking condition, we propose

$$f(\eta) = e^{-\eta} \qquad (V, 78)$$

With this form $f(\eta)$, the function $F(Pr)$ can be represented by the interpolation formula

$$F(Pr) = \frac{Pr}{\ln Pr - 0.4 + \frac{1.4}{Pr}} \qquad (V, 93)$$

The value of $G = F(1)$ is equal to unity.

For the dimensionless width of the linking zone, the formula of Nikuradze (V, 64), independently of the particular form of the linking function, leads to the value $\frac{1}{\kappa} = 2.5$.

For the determination of the numerical value of the dimensionless thickness of the laminar sublayer L, we can make use of experimental data from three sources:

1. From the experimental data of Nikuradze on velocity distribution: Combining (V, 80) with (V, 64) we get $L = 6.8$.

2. From our experimental data on the kinetics of dissolution at high viscosities of the solution: The approximate form of formula (V, 91) at high values of the Prandtl criterion will be

$$\frac{1}{M} \cong L \sqrt{\frac{2}{f}} \ Pr \qquad\qquad (V, 94)$$

Combination of this formula with our experimental data on turbulent dissolution (cf. Chapter II, Table 2) gives values of L between 2.5 and 4.6.

3. From literature data on the dependence of the heat transfer on the value of the Prandtl criterion: These data are usually represented in the form of empirical formulas, built after the type of Prandtl's formula, with empirically chosen expressions for φ as a function of the Reynolds and Prandtl criteria. Combining (V, 93) with Prandtl's formula, we conclude that our formula can be reduced to a form analogous to Prandtl's formula, if one puts in the latter

$$\varphi = L \sqrt{\frac{f}{2}} \left[1 + \frac{1}{\kappa L} \frac{F(Pr) - G}{Pr - 1} \right] \qquad (V, 95)$$

Combination of this expression with the empirical formulas for φ and with Blasius' law leads to the value L = 4.0.

Comparison of the Different Assumptions on the Velocity Distribution

Figure 18 compares the different assumptions on the velocity distribution on which different authors have based the calculation of processes of convective diffusion and heat transfer.

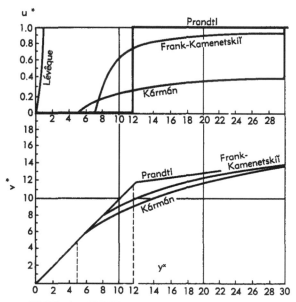

FIGURE 18.   COMPARISON OF DIFFERENT ASSUMPTIONS ON THE VELOCITY DISTRIBUTION IN A TURBULENT FLOW

Ordinates: for the top graph, the effective pulsation velocity; for the bottom graph, velocity of the flow. Abscissas: dimensionless distance from the wall. The scale of velocities is the magnitude

$$u_0 = V \sqrt{\tfrac{f}{2}}$$

the scale of length is the magnitude $\delta' = \dfrac{\upsilon}{u_0}$.

The curves are drawn with the aid of the formulas proposed by the different authors.

The abscissas are the dimensionless distances from the surface which, according to (V, 5), is defined by

$$y^* = \frac{y}{\delta'} = \frac{u_0}{\upsilon}\, y \qquad\qquad (V,\ 96)$$

The ordinates represent the effective pulsation velocities
and the tangential flow velocities, relative to the
magnitude

$$u_0 = V \sqrt{\tfrac{f}{2}}$$

where V is the velocity of the main flow at infinite distances from the surface. The curves correspond to the
formulas of Levich, Prandtl, von Karman, and the author.
The top graph gives the distribution of the effective pulsation velocity, the bottom graph the distribution of the
tangential velocity (i.e., the mean velocity of the flow).
The scale of velocity is the magnitude

$$u_0 = V \sqrt{\tfrac{f}{2}}$$

the scale of length is δ'. By effective pulsation velocity
we mean the magnitude which gives the correct value of the
turbulent exchange coefficient, if the mixing length is
expressed by formula (V, 67).

Within the limits of the laminar sublayer insofar as it has real physical existence, the effective pulsation velocity is zero, and the tangential velocity depends
on the distance according to (V, 75). After integration,
and passing to dimensionless coordinates, we get

$$v^* = y^* \qquad\qquad (V, 97)$$

The full line in the region immediately adjacent to the
surface has been drawn according to this formula. If the
laminar sublayer has no real physical existence, and the
normal component of the velocity vanishes only at the very
surface, then, according to Landau and Levich, the normal
component of the velocity in immediate vicinity of the
surface should increase proportionally to y^2. The proportionality coefficient we can indicate only for the

laminar boundary layer where, according to Levich

$$u* = \frac{u}{u_o} = y*^2 \qquad\qquad (V, 98)$$

$$v* = \frac{v}{u_o^{'}} = \int \frac{dy*}{1 + \kappa y*^2} \qquad\qquad (V, 99)$$

these formulas being valid only in immediate vicinity of the surface.

Formula (V, 98) is represented in Figure 18.

Formula (V, 99) at small $y*$ is practically not different from the straight line (V, 97); at high $y*$, formular (V, 99) loses its meaning, as the assumptions on which it was derived lose their validity. Thus, one cannot use Levich's method to derive the complete velocity distribution curve. This method can only give the limiting law of convection for high values of the Prandtl criterion. Comparison of formulas (V, 6) and (V, 30) shows that in the Landau and Levich theory, the magnitude

$$n \frac{1}{\sqrt{Pr}}$$

plays the role of the dimensionless thickness L of the laminar sublayer. At higher values of the Prandtl criterion, this magnitude should become much smaller than unity.

At a distance from the surface, in the core of the flow, all formulas should go over into the formula of Nikuradze (V, 64) which rests on reliable direct experimental material; the corresponding formula for the pulsation velocity is (V, 68).

In Prandtl's model, the pulsation velocity changes abruptly from zero to the value (V, 68) at the outer boundary of the laminar sublayer.

In order to obtain Nikuradze's formula for the core of the flow, one must assume that this abrupt change occurs at the value $y* = 11.7$ which has been calculated above (V, 74). Naturally, Prandtl's model, in which the linking zone does not figure at all, requires too high a value of the thickness of the laminar sublayer to give a correct velocity distribution.

In von Karman's model the laminar sublayer ends at the value $y* = 5$. From $y* = 5$ to $y* = 30$, there is an intermediate zone in which the velocity distribution is expressed by the formulas

$$u* = \frac{1}{2} - \frac{2.5}{y*} \qquad\qquad (V,\ 100)$$

$$v* = 5(1 + \ln \frac{y*}{5} \qquad\qquad (V,\ 101)$$

Direct experimental data for convection processes at higher values of the Prandtl criterion lead, as has been pointed out, to values of the dimensionless thickness L of the laminar sublayer of about 4.

We thus come to the conclusion that the existing experimental data can be described with the aid of the concept of the existence of a laminar sublayer of a dimensionless thickness of about 4, in which turbulent transfer is entirely absent.

However, there is no room yet for the conclusion that turbulence vanishes altogether at a finite distance from the surface. We can only say that close to the surface, there exists a sublayer of finite thickness, in which convective transfer is practically absent. The problem of the nature of the motion in this sublayer is a matter for further investigation.

Diffusion to Suspended Particles and
Local Structure of Turbulence

Up to this point we have discussed processes
linked with diffusion from the flow to an immobile solid
surface. A second important problem of chemical hydrodynamics is the investigation of processes linked with
diffusion to the surface of suspended particles, moving
under the action of turbulent motion of a gas or liquid.

This problem is encountered in the study of
microheterogeneous processes, the formal diffusional
kinetics of which have been discussed in Chapter II.
There the diffusion process to the surface of each individual suspended particle had been described by the diffusion velocity constant β, without any detailed consideration of the factors determining the value of that
magnitude. We now shall set ourselves the task of calculating the velocity of diffusion to the surface of suspended particles in turbulent motion. This problem has a
primary significance for such processes as the evaporation of liquid drops in a turbulent flow or the dissolution of suspended particles in stirring with an agitator.
The latter process is widely used in hydrometallurgy.

If the dimensions of the particles are very
small, their motion follows entirely the turbulent pulsations. The rate of the motion of the gas or liquid relative to the surface is in this case zero, and the diffusion
process obeys the laws of diffusion in a motionless medium.

Assuming the particles in a first approximation
to be of spherical shape, the value of the Nusselt criterion can be taken equal to 2, as was done in Chapter II.

Such an approach is entirely legitimate in application to very small particles, for example, to such
important instances of microheterogeneous processes as the
dissolution of colloidal sols or the catalytic hydrogenation on suspended catalysts.

With the increase of the size of the particles, they cease to follow the turbulent pulsation. The velocity of the gas or liquid relative to the surface of the particles becomes different from zero, and turbulent transfer begins to play an ever increasing role in the diffusion process. Very large particles are practically not carried along at all by the flow, and are subject to the same laws as a motionless surface placed in a flow.

How and at what dimensions will a transition take place from the limiting case of "small" to the limiting case of "large" particles?

In principle, the way to solve this problem is indicated by the theory of local structure of the turbulence developed in recent years by Kolmogorov and Landau. We shall use this theory in the form in which it is given in the book of Landau and Lifshits[5].

The mean relative velocity of motion of two particles of liquid or gas at a distance x from each other is expressed by

$$v_x = \sqrt[3]{\frac{\epsilon}{\rho} x} \qquad (V, 102)$$

where ρ is the density of the liquid or gas, and ϵ the dissipation energy, i.e., the energy dissipated (converted into heat as a result of viscosity) per unit volume in unit time.

The dissipation energy can be expressed by the tangential stress τ or the resistance coefficient f. The total energy dissipated over the whole volume w is equal to the work of the viscosity forces on the friction surface σ,

$$\epsilon w = \tau_0 \sigma V \qquad (V, 103)$$

where V is the velocity of the main flow or the linear

rate of revolution of the agitator.

Expressing, as usual, the tangential stress by the resistance coefficient, we get

$$\epsilon = \frac{f}{2} \rho V^3 \frac{\sigma}{\omega} \qquad\qquad (V,\ 104)$$

The ratio of the total volume in which the turbulent motion takes place and the friction surface will be called the "hydraulic radius" and will be designated by r'

$$r' = \frac{\omega}{\sigma} \qquad\qquad (V,\ 105)$$

In the particular case of a tube or channel this will co-incide with the usual definition of the hydraulic radius as the ratio of area of the cross-section and the wetted perimeter.

Substituting (V, 104) and (V, 105) in (V, 102), we get for the relative velocity of motion of two particles of liquid or gas at a distance x from one another, the expression

$$v_x = V \sqrt[3]{\frac{f}{2} \frac{x}{r'}} \qquad\qquad (V,\ 106)$$

The application of this theory to the calculation of processes of diffusion to the surface of particles sus-pended in a turbulent flow was discussed in a recently pub-lished article of Tunitskii[10]. He assumes that the sus-pended particle is completely carried along in its motion by the turbulent pulsations (which apparently is justified only for the particular case when the density of the par-ticle is not different from the density of the medium). Then, the relative velocity of flow at a distance x from the surface of the particle should be given directly by formula (V, 106). Assuming further (which can hardly be considered sufficiently substantiated) that the mixing

length is of the order of the distance x, Tunitskii gets
for the turbulent exchange coefficient the expression

$$A \approx V \sqrt[3]{\frac{f}{2r'}} \, x^{\frac{4}{3}} \qquad\qquad (V, \, 107)$$

He further introduces the thickness of the diffusion layer
δ_D, defined as the distance at which the turbulent exchange coefficient becomes of the same order of magnitude
as the molecular diffusion coefficient D. From formula
(V, 107) we get for the thickness of the diffusion layer

$$\delta_D = (\frac{D}{V})^{\frac{3}{4}} \, \sqrt[4]{\frac{f}{2r'}}, \qquad\qquad (V, \, 108)$$

If the dimension of the disperse particle is
smaller than the thickness of the diffusion layer, one
can disregard the influence of turbulence on the diffusion
process. In that case, diffusion to the surface of the
particle will take place as in a motionless medium. If,
on the contrary, the dimension of the disperse particle
is much greater than the thickness of the diffusion layer,
turbulent diffusion becomes essential. In that case, the
role of the reduced film at the surface of the particle
will be played by the diffusion layer, the thickness of
which is expressed by formula (V, 108).

It would be more correct not to make the arbitrary assumption that the particle is completely carried
along by the turbulent pulsations, but to calculate the
scale of the pulsations which are able to carry the particle along. This is done with the aid of the condition
of equality of the force acting on the particle on the
part of the flow and the force of inertia. The first is
expressed by

$$C_f \rho' v^2 4\pi r^2$$

and the second by

$$\frac{4}{3} \pi \rho r^3 \frac{v}{\tau}$$

where ρ is the density of the substance of the particle,
r its radius, ρ' the density of the medium (liquid or
gas), v the velocity, C_f the resistance coefficient
(of the order of unity), and τ the period of the motion,
equal to $\frac{\lambda}{v}$, where λ is the scale of the motion.

Hence, the maximum scale of motion not carrying
the particle along will be expressed by

$$\lambda \approx \frac{\rho}{\rho'} x \qquad\qquad (V, 109)$$

As the liquid flows around the particle at a distance λ
from its surface, the relative velocity has to be calculated by means of formula (V, 106) with the expression
(V, 109) for λ substituted instead of x. We then find
for the velocity of the flow relative to the surface of
the suspended particle

$$v = V \sqrt[3]{\frac{f}{2} \frac{r}{r'} \frac{\rho}{\rho'}} \qquad\qquad (V, 110)$$

where r is the radius of the particle, V the velocity
of the main flow, r' its scale (hydraulic radius), ρ
the density of the particle, and ρ' the density of the
medium.

The diffusion flow to the surface of the particle
can be found by the usual formulas of the theory of convective diffusion, only with the relative velocity (V, 110)
substituted for the flow velocity.

There has been no experimental verification so
far of these concepts. It would be extremely interesting
to study from this point of view such processes as the

evaporation of drops of the dissolution of disperse particles in the presence of strong artificial agitation.

Literature

1.   BUBEN and FRANK-KAMENETSKII, Zhur. Fiz. Khim. 20,
      225 (1946).

2.   HOFMANN, Z. ges. Kälteind. 44, 99 (1937); Forsch.
      Geb. Ingenieurwes 11, 159 (1940).

3.   EAGLE and FERGUSON, Proc. Roy. Soc. A 127 (1930).

4.   KARMAN, in "Problemy turbulentnosti" (Problems of
      turbulence), ed. Velikanov and Shveikovskii,
      p. 49 ff., Moscow 1936; Trans. Am. Soc. Mech.
      E Eng. 61, 704 (1939).

5.   LANDAU and LIFSHITS, Mekhanika Sploshnykh tel
      (Mechanics of continuous bodies) p. 111,
      Moscow 1944.

6.   LEVICH, Zhur. Fiz. Khim. 18, 335 (1944).

7.   PRANDTL, Phys. Z. 11, 1072 (1910); 29, 487 (1928).

8.   TEN BOSCH, Die Wärmeübertragung, 3rd ed., p. 119
      (1936).

9.   KUPRIJANOFF, Z. Tech. Phys. 16, 13 (1935).

10. TUNITSKII, Zhur. Fiz. Khim. 20, 1137 (1946).

11. MATTIOLI, Forschung 11, 149 (1940).

CHAPTER VI: THEORY OF COMBUSTION FROM THE POINT OF VIEW OF THE SIMILITUDE THEORY

Combustion is a chemical reaction under conditions of progressive self-acceleration, due to accumulation in the system of heat or of catalyzing active intermediate products. In the first case, the combustion is termed thermal, in the second, diffusional or chain combustion. Thermal combustion can be observed in any exothermal reaction if its rate increases rapidly with rising temperature. Diffusional combustion is possible only in the case of autocatalytic reactions.

The subject-matter of the theory of combustion is the simultaneous discussion and solution of the equations of chemical kinetics and the equations of heat transfer in the case of thermal, and of diffusion in the case of diffusional combustion.

The task of the present chapter is the application of the theory of similitude to the equations of the theory of combustion, as a combination of equations of chemical kinetics and of heat transfer. We shall endeavor to isolate what is general for all combustion phenomena and, at the same time, specific for combustion. Such a general and specific feature is the sharp exponential acceleration of the chemical reaction with rising temperature, in the presence of evolution of heat by the reaction itself, resulting in progressive rise of the temperature. We do not

aim in this book at a detailed mechanism of the combustion
reactions, and we shall take into consideration only the
more general features of the kinetics, in particular the
above-mentioned exponential dependence of the reaction
rate on the temperature. With the factors of heat transfer, diffusion, hydrodynamics, combined with chemical
kinetics, this will permit disclosure not only of the
basic qualitative features of the phenomena, but will also
give a quantitative theory describing correctly a number
of experimental facts.

Theory of Thermal Ignition

The theory of thermal ignition deals with the
question of what happens to a combustible mixture if it is
placed in a vessel the walls of which are at a temperature
T_o. Under certain conditions, one will observe a rapid
growth of the temperature to a very high value, close to
the theoretical maximum temperature of explosion

$$T_m = T_o + \frac{Q}{c} \qquad\qquad (VI, 1)$$

where Q is the thermal effect of the reaction, and c
the heat capacity of the reacting mixture. Under other
conditions, on the contrary, only an insignificant rise
of the temperature will be observed, to a stationary level
not too different from T_o. This stationary temperature
rise will then remain almost constant until a considerable
part of the substance has reacted. The conditions under
which one range goes over into the other are termed the
critical conditions of ignition. The very existence of a
sharp transition is, of course, contingent on certain conditions which will be formulated further below.

We shall first consider the simplest limiting case
of purely conductive heat transfer which will take place at
small values of the Grashof criterion. For a rigorous
solution of the problem of thermal ignition, we must turn

to the equation of heat conductance with continuously distributed sources of heat (I, 51)

$$c\rho \, \frac{\partial T}{\partial t} = \text{div } \lambda \text{ grad } T + q'$$

where q' is the density of the sources.

Solution of this equation under boundary conditions involving a given constant temperature T_o at the
wall surface, will give the temperature distribution in the
vessel as a function of time. At the ignition limit, the
character of this dependence should change sharply; there
should be an abrupt transition from small stationary
temperature rise to large and rapid rise. However, owing
to the great mathematical difficulties of integration of
the partial differential equation (I, 51), this direct
method of solution has not been attempted by anyone. Instead, one resorts to two approximate methods which are
the object of the nonstationary and the stationary theory
of thermal explosion. In the stationary theory, only the
temperature distribution over the vessel is considered,
without its change with time being taken into account. In
the nonstationary theory, on the contrary, the spatial
temperature distribution is not taken into consideration,
but a mean temperature of the reacting mixture, assumed
to be equal at all points of the reaction vessel, is introduced and its dependence on the time is examined.

The basic idea of the theory of thermal ignition
is due to van't Hoff[1]. According to it, the condition of
thermal ignition consists in the impossibility of a thermal
equilibrium between the reacting system and the surrounding medium. The qualitative formulation of this condition
as contact between the curve of heat supply and the straight
line of heat removal was first given by Le Chatelier[2].
The mathematical formulation was given by Semenov[3] who
obtained an expression for the relation between the explosion parameters (temperature and pressure at the

explosion limit) which was later confirmed by Zagulin and a number of other investigators.

The nonstationary theory of thermal explosion was further developed by Todes[4] and Rice and his collaborators[5] and the nonstationary theory by the author of this book[6].

Stationary Theory of Thermal Explosion

The stationary theory deals with the stationary equation of heat conductance with continuously distributed sources of heat. Its solution gives the stationary temperature distribution in the reacting mixture. The conditions under which such a stationary distribution becomes impossible are the critical conditions of ignition.

The stationary equation is of the form

$$\text{div } \lambda \text{ grad } T = - q' \qquad \text{(VI, 2)}$$

or, neglecting the temperature dependence of the heat conductivity

$$\lambda \Delta T = - q' \qquad \text{(VI, 3)}$$

where Δ is the Laplace operator.

The density of the sources of heat q' is in this case the amount of heat evolved by the chemical reaction in unit volume per unit time, i.e., the product of the thermal effect and the rate of reaction. If the rate of reaction depends on the temperature in accordance with Arrhenius' law, it can be represented in the form

$$ze^{-\frac{E}{RT}}$$

where z is a magnitude depending on the pressure and the composition of the mixture, but not depending on the temperature in a first approximation.

The density of the sources of heat will then be

expressed by

$$q' = Qze^{-\dfrac{E}{RT}} \qquad\qquad (VI, 4)$$

where   Q   is the thermal effect of the reaction.

Equation (VI, 2) can now be put in the form

$$\Delta T = -\frac{Q}{\lambda} ze^{-\dfrac{E}{RT}} \qquad\qquad (VI, 5)$$

and the task consists in integrating the equation (VI, 5) under the boundary condition $T = T_0$ at the wall of the vessel.

To analyze equation (VI, 5) by the methods of the similitude theory it is necessary to choose an expression for the dimensionless temperature that would permit the most suitable transformation of Arrhenius' law. Owing to the temperature in this law not standing under a differential sign, but being included in a complex exponential function, the usual methods of transformation to dimensionless variables are in this case inapplicable. This problem of transforming Arrhenius' law to dimensionless variables is fundamental for all applications of the theory of similitude to kinetics at variable temperature, in particular to the theory of combustion.

In this case, there are two methods for defining the dimensionless temperature. The simplest one is immediately evident at a first glance at Arrhenius' law. It consists in taking for the dimensionless temperature the magnitude

$$u = \frac{RT}{E} \qquad\qquad (VI, 6)$$

For dimensionless coordinates we shall take $\xi = \frac{x}{r}$ where r is the radius of the vessel (spherical or cylindrical). Equation (VI, 5) will then take the form

$$\Delta_\xi u = -\frac{Q}{\lambda}\frac{R}{E} r^2 z e^{-\frac{1}{u}} \qquad (VI, 7)$$

and the boundary conditions will be, at $\xi = 1$,

$$u = u_o = \frac{RT_o}{E}$$

In this way, we have in the equation only one dimensionless parameter

$$v = \frac{Q}{\lambda}\frac{R}{E} r^2 z$$

and in the boundary conditions, a second dimensionless parameter

$$u_o = \frac{RT_o}{E}$$

A solution of equation (VI, 7) satisfying the boundary conditions should be of the form

$$u = f(\xi, v, u_o) \qquad (VI, 8)$$

giving the dimensionless temperature u as a function of the dimensionless coordinates ξ with two dimensionless parameters v and u_o. The conditions under which such a stationary temperature distribution becomes impossible should be of the form

$$v = f(u_o) \qquad (VI, 9)$$

as neither the equation nor the boundary conditions contain any parameters other than v and u_o.

Finding of the analytical form of the function f in (VI, 9) was attempted by Todes and Kontorova[7].

Formula (VI, 9) represents the most general solution of the problem of thermal ignition in purely conductive heat exchange from the point of view of the

similitude theory. Of the greatest practical importance
is the fact that in all real cases one encounters values

$$u_o = \frac{RT_o}{E} \ll 1$$

It therefore appears reasonable to seek a limiting form
of the expression (VI, 9) at $u_o \longrightarrow 0$, and if such a
form exists, to utilize it in all practical problems.
Moreover, it develops that if one considers only the
limiting case $u_o \ll 1$, one not only obtains much simpler
and clearer results, but also all the specific features
proper to combustion phenomena stand out more distinctly.

In order to pass to this limiting case, let us
keep in mind that we seek the stationary temperature distribution below the explosion limit where the temperature
rise is small. We introduce the variable $\vartheta = T - T_o$ and
we assume $\vartheta \ll T_o$. The equivalence of this assumption
with $u_o \ll 1$ will be demonstrated post factum. We now
can write

$$e^{-\frac{E}{RT}} = e^{-\frac{E}{R(T_o + \vartheta)}} = e^{-\frac{E}{RT_o}\left(\frac{1}{1 + \frac{\vartheta}{T_o}}\right)}$$

Developing

$$\frac{1}{1 + \frac{\vartheta}{T_o}}$$

into a geometric series and omitting, by virtue of our
assumption $\vartheta \ll T_o$, all terms with powers of $\frac{\vartheta}{T_o}$ higher
than one, we get

$$e^{-\frac{E}{RT}} \cong e^{-\frac{E}{RT_o}\left(1 - \frac{\vartheta}{T_o}\right)} = e^{-\frac{E}{RT_o}} e^{\frac{E}{RT_o^2}\vartheta} \qquad (VI, 10)$$

Substituting for $e^{-\frac{E}{RT}}$ in (VI, 5) its approximate value (VI, 10) we get

$$\Delta \vartheta = \frac{Q}{\lambda} ze^{-\frac{E}{RT_o}} e^{\frac{E}{RT_o^2}\vartheta} \qquad (VI, 11)$$

as ϑ and T differ only by the constant additive T_o. Equation (VI, 11) has to be integrated under the boundary condition $\theta = 0$ at the wall of the vessel. The form of this equation shows us immediately which magnitude ought to be taken for the dimensionless temperature. It is the variable standing in the exponent, which we shall designate by θ

$$\theta = \frac{E}{RT_o^2} \vartheta \qquad (VI, 12)$$

Transforming equation (VI, 11) to the dimensionless variables θ and ξ, we get

$$\Delta_\xi \theta = -\frac{Q}{\lambda} \frac{E}{RT_o^2} r^2 ze^{-\frac{E}{RT_o}} e^{\theta} \qquad (VI, 13)$$

and the boundary conditions, at $\xi = 1$, $\theta = 0$. Now both the equation and the boundary conditions contain only one dimensionless parameter, which we shall designate by δ

$$\delta = \frac{Q}{\lambda} \frac{E}{RT_o^2} r^2 ze^{-\frac{E}{RT_o}} \qquad (VI, 14)$$

In this parameter, all the magnitudes characterizing the properties of the mixture and of the vessel, and which are essential for the thermal ignition, are assembled.

The solution of equation (VI, 13), representing the stationary temperature distribution, should be of the form

$$\theta = f(\xi, \delta) \qquad (VI, 15)$$

with one parameter δ, and the condition under which
such a stationary temperature distribution ceases to be
possible, i.e., the critical condition of ignition, is of
the form

$$\delta = const. = \delta_{cr} \qquad\qquad (VI, 16)$$

as neither the equation nor the boundary conditions contain any parameters other than δ. If the conditions of
the experiments, on substitution in equation (VI, 14),
give for δ a value less than the critical, a stationary
temperature distribution should establish itself. In the
contrary case, explosion will ensue.

The critical value of δ depends on the geometric shape of the vessel and can be found by numerical
integration of equation (VI, 13), as will be done in the
following chapter.

For a spherical vessel, $\delta_{cr} = 3.32$; for an infinitely long cylindrical vessel, $\delta_{cr} = 2.00$. The expression obtained for the critical condition of ignition
is in good agreement with experimental data for reactions
with known kinetics.

From the solution of equation (VI, 15) it is
seen that the maximum temperature rise below the explosion
limit is

$$\vartheta_{max} = (T - T_0)_{max} = \frac{RT_0^2}{E} f(0, \delta_{cr})$$

i.e., is proportional to $\dfrac{RT_0^2}{E}$. In this way, at
$RT_0 \ll E$, we shall have below the explosion limit,
$\vartheta \ll T_0$, i.e., the assumption made in the derivation of
the expression (VI, 10) is actually equivalent to a
limiting case $u_0 \ll 1$.

Thus, there actually exists a limiting form of
the expression (VI, 9) at $u_0 \longrightarrow 0$. Combining (VI, 9)

with (VI, 16), we come to the conclusion that at $u_0 \longrightarrow 0$, (VI, 10) should tend to the form

$$v = u_0^2 \, e^{\frac{1}{u_0}}$$

Only this limiting case is meaningful in the theory of combustion. If RT_0 is not small compared with E, one does not get the characteristic picture of the combustion phenomena; the general theory which discusses this case should be termed simply, theory of nonisothermal course of the chemical reaction, a limiting case of which is the theory of combustion.

From the point of view of the similitude theory, the main advantage gained by passing to that limiting case and defining the dimensionless termperature not as $\frac{RT}{E}$ but as $\frac{E}{RT_0^2} (T - T_0)$, consists in the fact that, whereas the equation itself contained only one dimensionless parameter in either case, the boundary conditions in the first case contained an additional dimensionless parameter u_0, but not in the second case. As a consequence, only one dimensionless parameter figures in the solution in the second case, as against two in the first case. The task of the analytical solution (in this instance, numerical integration) in the second case is reduced to the finding of only one constant number, the value of δ_{cr} for each given geometric shape of the vessel, whereas in the first case it would be necessary to find a function.

All these advantages are essentially rooted in the suitable choice of a yardstick of temperature. In the first case, as can be seen from formula (VI, 6), the role of a yardstick of temperature is displayed by the magnitude $\frac{E}{R}$. In all real problems, this yardstick is very large (of the order of 10,000°) compared with the temperature differences encountered. In contrast, in the

second case, as can be seen from formula (VI, 12), the
yardstick is the magnitude $\dfrac{RT_o^2}{E}$ which is evidently a
natural yardstick for the actual temperature differences,
the maximum temperature difference possible in stationary
temperature distribution being just of that order of
magnitude.

The purely conductive theory here disucssed can
be applied at sufficiently low pressures and small dimensions of the vessel when the influence of convection
can be disregarded. Under conditions where free convection becomes essential, the hydrodynamic equation, for the
case of free convection in gases, is added to the above
discussed equation of heat conductance (supplemented by
the term V grad T). This system of equations[*] which in
the approximation usually applied in the theory of free
convection is written as

$$a\Delta T = -\frac{Q}{c\rho}v + V\,\text{grad}\,T$$

$$(V\,\text{grad}\,V) = g\frac{\vartheta}{T} + \upsilon\Delta V$$

(where V is the velocity of convective motion of the
gas, v the rate of the chemical reaction, g the
gravity acceleration, υ the kinematic viscosity) has to
be solved with the velocity V eliminated. Velocity V
does not enter the boundary conditions and is not a determining parameter in the case of free convection.

Analysis of this system by the methods of the
theory of similitude shows that the critical value of the
parameter δ should become a function of the Grashof
criterion which in the given instance will be written

$$Gr = \frac{gd^3}{a^2}\frac{RT_o}{E}$$

<hr>

[*] The first equation is scalar, the second vectorial.

The condition of ignition in this case takes the form

$$\delta_{cr} = f(Gr) \qquad (VI, 17)$$

The form of this function should be universal for all cases and can be determined through evaluation of experimental data. At values of $Gr < 10,000$, one can disregard the influence of the convection. All available experimental data on thermal ignition for reactions with known kinetics refer to conditions where $Gr < 10,000$. Consequently, it is not possible yet to establish the form of the function in equation (VI, 17). One can obtain an approximate expression by the use of the nonstationary theory of the thermal explosion and the experimentally determined values of the heat exchange coefficient.

In the stationary theory, we disregard the change of the rate of reaction as a result of loss of reactant through combustion in the induction period. Instead of $ze^{-\frac{E}{RT}}$ one substitutes, in the case of a reaction with nonautocatalytic kinetics, the rate of the reaction at initial concentrations of the reactants, and in the case of autocatalytic reactions, the maximum rate of the reaction. The question of when such a simplification is legitimate, and all questions linked with the induction period, can be solved only by the nonstationary theory.

Nonstationary Theory of Thermal Explosion

The nonstationary theory deals with the thermal balance of the whole reaction vessel, assuming temperature to be the same at all its points. This assumption is admittedly incorrect in the conduction range, where the temperature gradient is by no means localized at the wall. It is, however, equivalent to a replacement of the mean values of all temperature-dependent magnitudes by their values at a mean temperature, and involves a relatively minor error.

If the volume of the reaction vessel is designated by ω, its wall surface area by S, and the heat exchange coefficient by α, the amount of heat evolved over the whole volume per unit time by the chemical reaction is equal to

$$\omega Q z e^{-\frac{E}{RT}}$$

and the amount of heat carried away to the wall

$$\alpha S(T - T_o)$$

The difference between these two amounts of heat, spent on raising the temperature of the gas per unit time, is

$$c\rho\omega \frac{dT}{dt}$$

where c is the heat capacity of the gaseous mixture and ρ its density (number of moles per unit volume).

Combining these expressions, we get the equation of thermal balance

$$c\rho \frac{dT}{dt} = \omega Q z e^{-\frac{E}{RT}} - \alpha S(T - T_o)$$

or

$$\frac{dT}{dt} = \frac{Q}{c\rho} z e^{-\frac{E}{RT}} - \frac{\alpha S}{c\rho\omega}(T - T_o) \qquad (VI, 18)$$

Transforming the exponent according to (VI, 10) and passing to the dimensionless temperature according to (VI, 12), we put the last equation in the form

$$\frac{d\theta}{dt} = \frac{Q}{c\rho} \frac{E}{RT_o^2} z e^{-\frac{E}{RT_o}} e^{\theta} - \frac{\alpha S}{c\rho\omega} \theta \qquad (VI, 18a)$$

with the initial condition $\theta = 0$ at $t = 0$.

Equation (VI, 18a) is not dimensionless. Each
of its members has the dimension of a reciprocal time.
In order to put it in a dimensionless form, it is necessary to introduce a natural yardstick of time. Equation (VI, 18) contains two magnitudes which can serve as
a yardstick

$$\tau_1 = \left(\frac{Q}{c} \frac{E}{RT_0^2} \frac{z}{\rho} e^{-\frac{E}{RT_0}} \right)^{-1} \qquad \text{(VI, 19)}$$

$$\tau_2 = \left(\frac{\alpha S}{c \rho \omega} \right)^{-1}$$

Consequently, the solution of (VI, 18) should be of the
form

$$\theta = f \left(\frac{t}{\tau} , \frac{\tau_2}{\tau_1} \right) \qquad \text{(VI, 20)}$$

where τ stands for either τ_1 or τ_2.

Thus, the dependence of the dimensionless
temperature on the dimensionless time contains one
dimensionless parameter $\frac{\tau_2}{\tau_1}$. The value of the parameter
determines the course of the curve of temperature as a
function of time. In particular, a sharp change of the
form of that curve should occur, if at all, at a definite
critical value of the parameter $\frac{\tau_2}{\tau_1}$, i.e., the critical
condition of ignition should be of the form

$$\frac{\tau_2}{\tau_1} = \text{const.} \qquad \text{(VI, 21)}$$

This result was first obtained by Todes[4]. Taking the values of α from the theory of heat transfer

$$\alpha = \text{Nu} \frac{\lambda}{d}$$

where Nu is a constant magnitude in the case of pure
conduction (and a function of the Grashof criterion in
the case of convection), one sees easily that the para-
meter $\dfrac{\tau_2}{\tau_1}$ coincides with the parameter δ introduced in
the stationary theory, except for a constant factor de-
pending on the geometric shape of the vessel.

Thus, either theory leads to the same form of
the critical condition of ignition. The stationary theory
permits more accurate determinations of the numerical value of the parameter in the conduction range, as it takes
into account the temperature distribution in the reaction vessel.

We shall elucidate the physical meaning of the
magnitudes τ_1 and τ_2. Equation (VI, 19) can be written
in the form

$$\frac{d\theta}{dt} = \frac{e^{\theta}}{\tau_1} - \frac{\theta}{\tau_2} \qquad\qquad \text{(VI, 22)}$$

In the right-hand member, the first term is proportional
to the amount of heat evolved by the reaction, the second
term, to the amount of heat carried away to the wall. Far
in the ignition range, the first term will be much greater than the second; under these conditions, one can disregard the removal of heat and view the thermal explosion
as adiabatic. For an adiabatic thermal explosion, the
time dependence of the temperature should be of the form

$$\theta = f\left(\frac{t}{\tau_1}\right) \qquad\qquad \text{(VI, 23)}$$

The time within which a given value of θ is attained is
under these conditions proportional to the magnitude τ_1.
Consequently, the induction period, i.e., the time within
which the explosion will occur, is, in the instance of
adiabatic thermal explosion, proportional to τ_1. As the

analytical solution of Todes[4] shows, the proportionality
factor in this case, too, is equal to unity. Therefore,
the magnitude τ_1 is termed the adiabatic induction
period. Conversely, if there were no heat evolved as a
result of the reaction, only the second term would remain
in the right-hand member of equation (VI, 22), and the
equation would describe the process of cooling of the gas
in the given vessel. Clearly, for such a process the
natural yardstick of time would be the magnitude τ_2
which, therefore, is termed the characteristic time of
heat exchange. The condition (VI, 21) can now be formulated as follows: at the ignition limit, the ratio of the
characteristic time of heat exchange and the adiabatic
induction period is constant.

With the aid of the results obtained, we can
point out the conditions under which it is legitimate to
disregard the loss of reactant through combustion during
the induction period. To this end, we shall calculate
the isothermal combustion of the reactant during the
adiabatic induction period,

$$\frac{ze^{-\frac{E}{RT_0}}\tau_1}{\rho} = \left(\frac{Q}{c}\frac{E}{RT_0^2}\right)^{-1} = \frac{1}{B} \qquad (VI, 24)$$

where B is a new dimensionless parameter

$$B = \frac{Q}{c} \cdot \frac{E}{RT_0^2} \qquad (VI, 25)$$

Combining the definition of the dimensionless temperature
(VI, 12) with the expression for the maximum explosion
temperature, we see that the parameter B is nothing but
the value of the dimensionless temperature θ at the
maximum explosion temperature

$$B = \theta_m = \frac{E}{RT_o^2} (T_m - T_o)$$

where $T_m = \frac{Q}{c} + T_o$ is the theoretical maximum temperature of explosion, calculated on the assumption of a constant heat capacity.

If that parameter is large as compared with unity, it is entirely permissible to disregard the loss of reactant through combustion over the induction period of the thermal explosion far from the limit (where that period is close to adiabatic). On the other hand, if the value of B is not large enough, there will be no explosion at all. The maximum temperature will differ little from the stationary temperature rise, and there will be no sharp transition from one range to the other.

Thus, the approximations in which all the characteristic features of the combustion phenomena find their expression, become legitimate under the following two conditions:

$$\frac{1}{u_o} = \frac{E}{RT_o} \gg 1$$

$$B = \frac{E}{RT_o} \frac{Q}{c} \gg 1$$

The theory of combustion is the limiting case of the theory of the nonisothermal course of a chemical reaction at large values of the two above dimensionless parameters.

Thermal Propagation of a Flame

In discussing the thermal propagation of a flame, we start from the same equation of heat conductance in a medium with continuously distributed sources of heat (I, 51), but with essentially different conditions of the problem. What is sought is a stationary state of propagation of the flame, i.e., a state in which the flame spreads at a constant linear rate relative to the reacting mixture. Inasmuch as only the flame velocity relative to the gas is significant, we can consider the case where the flame is motionless, and the gas is blown through it at a definite velocity. Except for such secondary effects as the influence of the wall on the hydrodynamic conditions in the flow, both cases are entirely equivalent to each other, and the velocity at which it is necessary to blow the gas for the flame to stay motionless is equal to the velocity of propagation of the flame relative to a motionless gas. We shall designate that velocity by w.

We introduce a coordinate system bound with the flame, and we designate the coordinates of that system by x, and the coordinates of the system bound with the gas by ξ. If the flame spreads in the motionless gas, the coordinate system x is mobile and the system ξ immobile. If, on the other hand, the gas is blown through a motionless flame, the system ξ is mobile and the system x immobile.

In the coordinate system ξ, the regimen is not stationary; in this system, equation (I, 51) is applicable. In the coordinate system x the regimen is, by convention, stationary, in this system $\frac{\partial T}{\partial t} = 0$. The heat conductance equation (I, 51) is not applicable in the coordinate system x, as the mass flow of the gas gives rise to an additional transport of heat not taken into account in equation (I, 51). Thus, in the system ξ

$$c\rho\left(\frac{\partial T}{\partial t}\right)_{\xi} = \text{div } \lambda \text{ grad}_{\xi} T + q'$$

The coordinates x and ξ are related by

$$x = \xi + wt$$

By the properties of partial derivatives,

$$\left(\frac{\partial T}{\partial t}\right)_\xi = \left(\frac{\partial T}{\partial t}\right)_x + \left(\frac{\partial T}{\partial x}\right)_t \left(\frac{\partial x}{\partial t}\right)_\xi = \left(\frac{\partial T}{\partial x}\right)_x + w\left(\frac{\partial T}{\partial x}\right)_t$$

and by the stationarity condition, $\left(\frac{\partial T}{\partial t}\right)_x = 0$. Hence, in the x system, equation (I, 51) takes the form

$$c\rho w\, \frac{\partial T}{\partial x} = \text{div } \lambda \text{ grad } T + q' \qquad\qquad (VI, 26)$$

We shall disregard the loss of heat at the wall. The problem thus becoming one-dimensional, with only one independent variable standing in the equation, the partial derivatives can be replaced by total derivatives.

In the following, as it is intended only to examine the problem by the methods of the similitude theory, without aiming at an accurate analytical solution, we shall disregard the temperature dependence of the heat conductivity although, for understandable reasons, this is less permissible in the given problem than in any other.

Equation (IV, 26) will then take the form

$$a\, \frac{d^2 T}{dx^2} - w\, \frac{dT}{dx} + \frac{q'}{c\rho} = 0 \qquad\qquad (VI, 27)$$

The magnitude q' can be represented by

$$q' = \frac{Q\rho}{\tau} \qquad\qquad (VI, 28)$$

where Q is the heat effect of the reaction, and τ the characteristic time of the reaction (a magnitude inversely proportional to the relative rate of reaction) i.e., the time within which, at a constant rate of reaction, the whole amount of the reactant present would burn away. Equation (VI, 27) then takes the form

$$a \frac{d^2T}{dx^2} - w \frac{dT}{dx} + \frac{Q}{c\tau} = 0 \qquad (VI, 29)$$

where τ is a function of the temperature, the form of which must be given.

Introducing the theoretical maximum temperature of combustion, calculated on the assumption of a constant heat capacity, $T_m^* = T_0 + \frac{Q}{c}$, where T_0 is the initial temperature of the combustible mixture, one gets the following form of equation (VI, 29):

$$a \frac{d^2T}{dx^2} - w \frac{dT}{dx} + \frac{T_m^* - T_0}{\tau} = 0 \qquad (VI, 30)$$

Equation (VI, 30) has to be integrated under the boundary conditions, at $x = -\infty$, $T = T_0$, and at $x = +\infty$, $T = T_m$ (T_m is the true maximum temperature; as we disregard the laws of heat at the wall, there is no reason for the burned gas to cool off).

From this boundary condition it follows auto- matically that at $x = \pm \infty$, the derivative $\frac{dT}{dx}$ will be zero.

Equation (VI, 30) is of the second order; its general integral contains two arbitrary constants. It might appear that, by a suitable choice of these constants, one should always be able to satisfy these two boundary conditions with any value of the parameter w. However, this is not so. The reason is that the boundary conditions are given only for infinite x. Neither the equation nor the boundary conditions contain the magnitude x, which is due to the obvious arbitrariness of the position of the origin of the coordinates. Consequently, one of the arbi- trary constants should enter the solution in the form x + c and thus falls out from the boundary conditions at $\pm \infty$. Therefore the boundary conditions can be satisfied only at a definite value of the parameter w which gives the want- ed velocity of stationary propagation of the flame. The same fact can also be expressed in this way: Inasmuch as the equation does not contain the magnitude x, one can apply to it the method of lowering of the order accepted

for such equations; one can take T as the independent
variable, and $y = \frac{dT}{dx}$ for the function wanted. One then
will get the nonlinear equation of the first order

$$ay \frac{dy}{dT} - wy + \frac{T_m^* - T_0}{\tau} = 0 \qquad (VI,\ 31)$$

with the boundary conditions, $y = 0$ at $T = T_0$ and at
$T = T_m$.

An equation of the first order can satisfy two
boundary conditions only at a definite value of the parameter w, which gives the wanted velocity of stationary
propagation of the flame.

As we have disregarded the temperature dependence
of the heat conductivity, it is natural to disregard also
the temperature dependence of the heat capacity and to put
$T_m^* = T_m$, which we shall do in the following. The results
obtained can then easily be made more accurate by introducing the temperature dependence of the physical constants,
but we shall not concern ourselves with it here.

In equation (VI, 30), the most essential magnitude
is τ, the reciprocal chemical reaction rate constant. It
is a function of the temperature on which the rate of the
chemical reaction depends according to Arrhenius' law. It
is clear from the foregoing that the problem of choice of a
natural yardstick of the temperature, very essential from
the point of view of the similitude theory, is linked with
the temperature dependence of the reaction rate.

The solution of the problem of the propagation of
the flame, with the temperature dependence of the rate of
reaction taken into account, was obtained only very recently.
In the previously current simplified theories, a physically
incorrect temperature dependence of the reaction rate was
assumed. The most perfected among them (from the formal
point of view) appears to be the theory of Daniell[8]. Even
though this theory is not correct for any real case, it is
of interest as the simplest mathematical model. We shall
therefore discuss it as an example, before taking up the
accurate theory of propagation of a flame.

Theory of Daniell

In Daniell's theory it is assumed that up to a certain temperature T_1, called the "ignition temperature". the rate of the reaction is equal to zero. Beginning from that temperature, the reaction either proceeds at a constant temperature-independent rate or takes place instantaneously after the lapse of a certain time, called the "induction period", also independent of the temperature. Both variants of the theory differ only in the form of the analytical solution but are identical from the point of view of the similitude theory.

Thus, in Daniell's theory, the temperature dependence of the rate of reaction is given by

$$\text{at} \quad T < T_1, \qquad \tau = \infty$$
$$\text{at} \quad T > T_1, \qquad \tau = \text{const.}$$

Correspondingly, equation (VI, 30) or (VI, 31) ought to be solved separately for each of these temperature zones. The solutions obtained should satisfy the continuity condition at $T = T_1$. It is easily seen that if one takes $T_1 - T_0$ for the yardstick of temperature, the linking conditions will contain no additonal parameters besides those following from the equation itself. Consequently, the magnitude $T_1 - T_0$ is the natural yardstick of temperature in Daniell's theory.

The problem now boils down to transforming equation (VI, 30) to dimensionless variables. Since we have disregarded the loss of heat at the wall, the conditions of the problem contain no magnitudes of the dimension of a length. We therefore shall introduce a conventional yardstick of length d which must subsequently be determined from the magnitudes entering the equation itself.

Introducing the dimensionless temperature

$$\theta = \frac{T - T_0}{T_1 - T_0}$$

and the dimensionless coordinate

$$\xi = \frac{x}{d}$$

we bring equation (VI, 30) to the form

$$\frac{a(T_1 - T_0)}{d^2} \cdot \frac{d^2\theta}{d\xi^2} - \frac{w(T_1 - T_0)}{d} \frac{d\theta}{d\xi} + \frac{T_m - T_0}{\tau} = 0$$

Dividing the whole equation by $\dfrac{a(T_1 - T_0)}{d^2}$

we get the dimensionless equation

$$\frac{d^2\theta}{d\xi^2} - \frac{wd}{a} \frac{d\theta}{d\xi} + \frac{T_m - T_0}{T_1 - T_0} \frac{d^2}{a\tau} = 0 \qquad\qquad (VI, 32)$$

The boundary conditions take the form

at $\xi = +\infty$, $\theta = \theta_m$;   at $\xi = -\infty$, $\theta = 0$

and the continuity condition at $\theta = 1$.

Thus we have in the equation two dimensionless parameters

$$\frac{wd}{a} \quad \text{and} \quad \frac{T_m - T_0}{T_1 - T_0} \frac{d^2}{a\tau}$$

and one additional parameter,

$$\theta_m = \frac{T_m - T_0}{T_1 - T_0}$$

in the boundary conditions. The first two parameters are not, however, determining; they include the magnitude d (the actual yardstick of length) which does rlot enter the conditions of the problem and must be determined. Elimination of the magnitude d from the first two parameters (through division of the second by the square of the first) gives the determining parameter

$$\frac{a}{w^2 \tau} \frac{T_m - T_o}{T_i - T_o}$$

Thus, the conditions of the problem contain two determining parameters,

$$\frac{a}{w^2 \tau} \frac{T_m - T_o}{T_i - T_o} \quad \text{and} \quad \frac{T_m - T_o}{T_i - T_o}$$

this system of dimensionless parameters is evidently equivalent to the system

$$\frac{a}{w^2 \tau}; \quad \frac{T_m - T_o}{T_i - T_o}$$

In the absence of other parameters in the conditions of the problem, the condition of existence of a stationary state of propagation, i.e., of a solution of equation (VI, 32) satisfying the boundary conditions, should have the form of a functional relation between these two parameters

$$\frac{a}{w^2 \tau} = f\left(\frac{T_m - T_o}{T_i - T_o}\right) \qquad (VI, 33)$$

whence the velocity of stationary propagation of the flame is

$$w = \sqrt{\frac{a}{\tau} F\left(\frac{T_m - T_o}{T_i - T_o}\right)} = \sqrt{\frac{a}{\tau} F(\theta_m)} \qquad (VI, 34)$$

This is Daniell's formula. For the function $F(\theta_m)$, analytical solution gives $F(\theta_m) = \theta_m - 1$ if one assumes that at $T > T_i$, the reaction proceeds at a constant temperature-independent rate, and $F(\theta_m) = \ln\theta_m$ if one assumes that the reaction takes place instantaneously after the lapse of a temperature-independent induction period.

The thickness of the flame zone is proportional to the natural yardstick of length

$$d \sim \frac{a}{w} \, \Phi \left(\frac{T_m - T_o}{T_i - T_o} \right) \sim \sqrt{a\tau} \; \varphi \left(\frac{T_m - T_o}{T_i - T_o} \right) \qquad (VI, \; 35)$$

Combining (VI, 34) with (VI, 35), we get

$$wd \sim aF' \; \frac{T_m - T_o}{T_i - T_o} \qquad (VI, \; 36)$$

This theory rests on assumptions about the kinetics of the reaction which do not correspond to any reality. Therefore, its results are not only unusable for quantitative computations but do not even give a qualitative agreement with the experiment. According to Daniell's theory, at an initial temperature of the combustible mixture equal to the ignition temperature, the flame velocity should become infinite, which is never observed in reality. The rate of a chemical reaction is a continuous function of the temperature, described by Arrhenius' law. The limit of thermal ignition, discussed in the foregoing paragraphs, has nothing to do with the concept of "ignition temperature" which is used in Daniell's theory.

Theory of Zeldovich[9]

A correct theory of propagation of the flame must take into account the true temperature and reactant-concentration dependence of the rate of the chemical reaction. Let the rate of reaction be expressed by

$$f(n)ze^{-\frac{E}{RT}}$$

where   n   is the relative reactant concentration, and

$f(n)$ so defined that $f(1) = 1$. Equation (VI, 27) will then take the form

$$a \frac{d^2T}{dx^2} - w \frac{dT}{dx} + \frac{Q}{c\rho} f(n)ze^{-\frac{E}{RT}} = 0 \qquad (VI, 37)$$

with the boundary conditions, at $x = -\infty$, $T = T_o$, and at $x = +\infty$, $T = T_m$.

The reactant concentration at the front of the flame will change not only as a result of the reaction itself but also as a result of diffusion. Therefore, along with the heat conductance equation (VI, 35), we shall have to consider also the equation of the diffusion which, when transformed for stationary flame propagation in the same way as was done above with the heat conductance equation, will take the form

$$D \frac{d^2n}{dx^2} - w \frac{dn}{dx} - \frac{1}{\rho} f(n)ze^{-\frac{E}{RT}} = 0 \qquad (VI, 38)$$

where D is the diffusion coefficient of the reactant in the reacting mixture. The boundary conditions are, at $x = -\infty$, $n = 1$, and at $x = +\infty$, $n = 0$.

We know that for gases the diffusion and the temperature conductivity coefficients are very close as to magnitude. If one disregards the difference, equations (VI, 37) and (VI, 38) will prove to be entirely similar to each other. By means of a simple transformation of the variables one can make them identical. To this end, we introduce in equation (VI, 37) the variable $\vartheta = T_m - T$; the equation then takes the form

$$- a \frac{d^2\vartheta}{dx^2} + w \frac{d\vartheta}{dx} + \frac{Q}{c\rho} f(n)ze^{-\frac{E}{R(T_m - \vartheta)}} = 0 \qquad (VI, 39)$$

The boundary conditions are, at $x = -\infty$, $\vartheta = T_m - T_o$, and at $x = +\infty$, $\vartheta = 0$. Multiplying equation (VI, 38) by $-\frac{Q}{c}$, we get

$$- D \frac{Q}{c} \frac{d^2 n}{dx^2} + w \frac{Q}{c} \frac{dn}{dx} + \frac{Q}{c\rho} f(n) z e^{-\frac{E}{R(T_m - \vartheta)}} = 0 \quad (VI, 40)$$

If one introduces in equation (VI, 40) a new variable $\eta = \frac{Q}{c} n$, we get evidently

$$- D \frac{d^2 \eta}{dx^2} + w \frac{d\eta}{dx} + \frac{Q}{c\rho} f(n) z e^{-\frac{E}{R(T_m - \vartheta)}} = 0 \quad (VI, 41)$$

which at $D = a$ is identical with (VI, 39).

The boundary conditions of equation (VI, 41) are, at $x = -\infty$, $\eta = \frac{Q}{c}$, and at $x = +\infty$, $\eta = 0$.

If, moreover, one disregards the temperature dependence of the heat capacity, $\frac{Q}{c} = T_m - T_o$, and the boundary conditions of equation (VI, 41) will coincide exactly with the boundary conditions of equation (VI, 39). Thus the magnitudes ϑ and η turn out to be solutions of identical differential equations with identical boundary conditions. Consequently, they must be identically equal to each other,

$$\eta = \vartheta \quad (VI, 42)$$

or

$$\frac{Q}{c} n = T_m - T \quad (VI, 42a)$$

or, because $\frac{Q}{c} = T_m - T_o$,

$$n = \frac{T_m - T}{T_m - T_o} \quad (VI, 43)$$

The result (VI, 43) can be formulated as similitude of the
fields of concentrations and of temperature. We have
derived it for the case where there is only one reactant
which determines the rate of reaction. It is easy to
generalize this result and extend it to the case where the
rate of the chemical reaction depends on the concentrations
of several reactants (with, of course, the diffusion coefficients taken to be the same and equal to the temperature conductivity coefficient of the mixture), on condition that their concentrations be linked by unambiguous
stoichiometric relations.

     The latter condition is always fulfilled for the
initial reactants; it is also fulfilled for the final
products in case the reaction is described by one stoichiometric equation but it is never fulfilled for the intermediate products.

     Thus, the similitude of the fields of concentra-
tions and of temperatures hinges on the condition that the
rate of the chemical reaction depends only on the concentrations of the substances bound by unambiguous stoichiometric relations. Moreover, we consider the diffusion coefficients of all these substances identical and equal to
the temperature conductivity coefficient of the mixture,
and we disregard the temperature dependence of the heat
capacity. Under these conditions and in this approximation, the system of equations (VI, 37) and (VI, 38) can
be reduced to one equation; substituting in (VI, 39) the
expression of η by ϑ from (VI, 42) we get the equation

$$- a \frac{d^2\vartheta}{dx^2} + w \frac{d\vartheta}{dx} + \frac{Q}{c\rho} f(\frac{c}{Q}\vartheta) \, ze^{-\frac{E}{R(T_m - \vartheta)}} = 0 \qquad (VI, 44)$$

which has to be integrated under the boundary conditions,
at $x = -\infty$, $\vartheta = T_m - T_0 = \frac{Q}{c}$, and at $x = +\infty$, $\vartheta = 0$.

     Because of the very strong exponential tempera-
ture dependence of the rate of reaction, this last term

of the left-hand member of equation (VI, 44) is significant only at temperatures close to T_m; at considerably
lower temperatures, i.e., at high values of ϑ , this
term becomes very small and can be disregarded. The very
phenomenon of stationary flame propagation is due just to
that; if the rate of reaction were not negligibly small
at the temperature T_o, the mixture would complete its
reaction spontaneously after the lapse of some time, independently of the propagation of the flame.

We therefore can resort to a development of the
Arrhenius law exponent in a series, as we have done in
the problem of the thermal explosion, putting $\vartheta \ll T_m$.
Only here one takes for the natural zero of temperature
not T_o but T_m.

Taking for the dimensionless temperature the
magnitude $\theta = \dfrac{E}{RT_m^2}\vartheta$, and introducing the dimensionless
coordinate $\xi = \dfrac{x}{d}$, where d is an arbitrary yardstick
of length, we bring equation (VI, 44) to the dimensionless
form

$$- \frac{d^2\theta}{d\xi^2} + \frac{wd}{a}\frac{d\theta}{d\xi}$$

$$+ \frac{Q}{c\rho}\frac{E}{RT_m^2}\frac{d^2}{a} f\left(\frac{c}{Q}\frac{RT_m^2}{E}\theta\right)ze^{-\frac{E}{RT_m}} \cdot e^{-\theta} = 0 \qquad \text{(VI, 45)}$$

with the boundary conditions, at $\xi = -\infty$,

$$\theta = \frac{Q}{c}\frac{E}{RT_m^2} = \frac{E}{RT_m^2}(T_m - T_o)$$

and at $\xi = +\infty$, $\theta = 0$. The equation and the boundary
conditions include three dimensionless parameters

$$A = \frac{wd}{a}$$

$$B = \frac{Q}{c} \frac{E}{RT_m^2} \frac{d^2}{a} \frac{z}{\rho} e^{-\frac{E}{RT_m}}$$

$$C = \frac{Q}{c} \frac{E}{RT_m^2} = \frac{E}{RT_m^2} (T_m - T_0)$$

The first two are not determining, as they contain the magnitude d which is not included in the conditions of the problem. Dividing B by A^2, we get the determining parameter

$$D = \frac{a}{w^2} \frac{Q}{c} \frac{E}{RT_m^2} \frac{z}{\rho} e^{-\frac{E}{RT_m}}$$

Solution of equation (VI, 45) will give the dependence of the dimensionless temperature θ on the dimensionless coordinate ξ with two parameters C and D. The condition that this solution satisfy the boundary conditions, containing only the parameter C, must be of the form

$$D = \Phi(C)$$

Hence, we get for the velocity of the stationary flame propagation

$$w = \sqrt{a \frac{z}{\rho} e^{-\frac{E}{RT_m}} \frac{Q}{c} \frac{E}{RT_m^2} \Phi\left(\frac{Q}{c} \frac{E}{RT_m^2}\right)} \qquad \text{(VI, 46)}$$

The magnitude

$$\frac{\rho}{z} e^{-\frac{E}{RT_m}}$$

is the time within which all the reactant would be consumed at the initial reaction rate and at the temperature T_m, i.e, it is the characteristic time of reaction at that temperature - we shall designate it by τ_m. The parameter

$$\frac{Q}{c}\frac{E}{RT_m^2} = \frac{E}{RT_m^2}(T_m - T_0)$$

is the value of the dimensionless temperature at the maximum flame temperature T_m; it is natural to designate it by θ_m. This parameter in the expression (VI, 46) stands under the function sign Φ and before that function; it is natural to combine these two factors and to designate $\theta_m\Phi(\theta_m)$ by a new function $F(\theta_m)$. We then get for the flame propagation velocity the following simple final expression

$$w = \sqrt{\frac{a}{\tau_m}F(\theta_m)} \qquad\qquad (VI, 47)$$

As we have shown in the discussion of the thermal explosion, the typical pattern of combustion phenomena is obtained when the reaction is sufficiently exothermal and its rate increases rapidly enough with the temperature. In this case, these conditions can be quantitatively formulated as

$$\frac{E}{RT_m} \gg 1$$

$$\theta_m \gg 1 \qquad\qquad (VI, 48)$$

Of the first condition, we have already made use in developing the exponent in a series. Thus, it has meaning to look for the limiting form of the function F in the expression (VI, 47) when also the second condition (VI, 48) is fulfilled. The method of analytic solution of the equation (VI, 45) for this limiting case was found by Zeldovich. This method gives the form of the function $F(\theta_m)$ for any form of the function $f(n)$ in equation (VI, 37), i.e., for any reactant-concentration dependence of the rate of reaction. In particular, in the simplest case when the kinetics of the reaction is given as $f(n) = n^p$, i.e., when the reaction is of a definite order p with respect to the reactant concentration, one gets for the function $F(\theta_m)$

$$F(\theta_m) = \frac{2p!}{\theta_m^{p+1}} \qquad\qquad (VI, 49)$$

Diffusional Flame Propagation

In the case of an autocatalytic reaction, the propagation of the flame can be due not only to a transfer of heat from the flame zone to the fresh gas but also to diffusion from that zone of catalytically active reaction products. If the temperature rise in the flame is very small (cool flames), the propagation will be entirely due to such a diffusion mechanism.

In order to find the velocity of diffusional propagation of the flame, we must solve the diffusion equation, transformed for the case of a stationary propagation of the flame in analogy to equation (VI, 38).

The rate of the reaction will be given by

$$\frac{dn}{dt} = \varphi\, f(n)$$

where n is the relative concentration of the catalytically active product, and φ the kinetic coefficient (self-acceleration coefficient), analogous to the rate constant.

The temperature dependence of the reaction rate will not be taken into account in this case; we take the temperature rise in the flame to be sufficiently small to be negligible (isothermal diffusional flame propagation). On these assumptions we get

$$D \frac{d^2 n}{dx^2} - w \frac{dn}{dx} + \varphi \, f(n) = 0 \qquad\qquad (VI,\ 50)$$

with the boundary conditions, at $x = -\infty$, $n = 0$; at $x = +\infty$, $n = 1$ in the case of autocatalysis by final products, and $n = 0$ in the case of autocatalysis by intermediate products.

As in the theory of thermal flame propagation, we consider here only the diffusion of one reactant.

This is not permissible in the case of autocatalysis by intermediate products, and it is necessary to solve a system of equations.

Since n is a dimensionless magnitude, we need to replace only x by the dimensionless coordinate $\xi = \frac{x}{d}$. After that, we get the dimensionless equation

$$\frac{D}{d^2} \frac{d^2 n}{d\xi^2} - \frac{w}{d} \frac{dn}{d\xi} + \varphi \, f(n) = 0$$

or

$$\frac{d^2 n}{d\xi^2} - \frac{wd}{D} \frac{dn}{d\xi} + \frac{\varphi d^2}{D} f(n) = 0 \qquad\qquad (VI,\ 51)$$

In this way, the equation contains two dimensionless parameters $\frac{wd}{D}$ and $\frac{\varphi d^2}{D}$. None of them is determining, as both include the magnitude d, which does not appear in the conditions of the problem. Eliminating this magnitude by dividing the second parameter by the square of the first, we get the determining parameter $\frac{\varphi D}{w^2}$.

On account of the absence of other parameters in both the equation and the boundary conditions, the condition of the stationary state takes the form

$$\frac{\varphi D}{w^2} = \text{const.}$$

whence the velocity of stationary flame propagation

$$w = A \sqrt{\varphi D} \qquad (VI, 52)$$

The value of the constant coefficient A depends on the
form of the function f(n); if this function has a complex form and contains any dimensionless parameters, the
latter will of necessity appear in the solution (VI, 52).

Literature

1.   VAN'T HOFF, Etudes de dynamique chimique, p. 161,
        Amsterdam, 1884.

2.   Quoted after JOUGUET, Mécanique des explosifs,
        p. 141. Paris, 1937. See also: TAFFANEL,
        Comptes rendus 156, 1544 (1913); 157, 469,
        595, 714 (1913).

3.   SEMENOV, Z. physik. Chem. 48, 571 (1928);
        Tsepnye reaktsii (Chain reactions), p. 116,
        Leningrad, 1934.

4.   TODES, Zhur. Fiz. Khim. 4, 71 (1933); 13, 868
        (1939); 13, 1594 (1939); 14, 1026 (1940);
        14, 1447 (1940); Acta physicochimica 5, 785
        (1936).

5.   RICE, J. Am. Chem. Soc. 57, 310, 1044, 2212
        (1935); J. Chem. Phys. 7, 701 (1939).

6.   FRANK-KAMENETSKII, Zhur. Fiz. Khim. 13, 738
        (1939); Acta physicochim. 10, 365 (1939);
        16, 357 (1942); 20, 729 (1945).

7.   TODES and KONTAROVA, Zhur. Fiz. Khim. 4, 81
        (1933).

8.   DANIELL, Proc. Roy. Soc. A 126, 393 (1930).

9.   ZELDOVICH and FRANK-KAMENETSKII, Zhur. Fiz.
        Khim. 12, 100 (1938).

CHAPTER VII: TEMPERATURE DISTRIBUTION IN THE REACTION
VESSEL AND STATIONARY THEORY OF THE
THERMAL EXPLOSION

Stationary Theory

In the last chapter, the general form of the
conditions of thermal ignition was obtained by the methods
of the similitude theory. We shall now examine this problem analytically in order to arrive at concrete numerical
results. To this end, it is necessary to solve the problem of the stationary temperature distribution in the
system where the chemical reaction takes its course.

The theory of thermal explosion, proposed by
Semenov[1], which has become the basis for all further work
in this field, rests on the assumption that the temperature can be taken to be equal at all points of the explosion vessel. This idea of "homogeneous ignition" is
not in accord with experimental facts; it is well known
that ignition always starts at one point, and then the
flame spreads over the vessel. As has been correctly remarked by Todes[2], the assumption of temperature equality
at all points of the vessel in the pre-explosion period
would be justified only in the presence of a convection so
intense that all of the temperature gradient would be at
the wall of the vessel. But under these conditions the
limit of thermal ignition should primarily depend on the
thickness and the material of the wall, which was not experimentally observed. If, on the contrary, one takes the

heat transfer within the gas to be purely conductive, one
will find a certain temperature distribution in the gaseous
mixture within the vessel; the temperature will be highest
in the center of the vessel and that is where the ignition
ought to start. The heat exchange coefficient and the
critical condition of ignition will be determined by this
temperature distribution. Ignition ought to occur when a
stationary temperature distribution becomes impossible,
as in the case of thermal breakdown of dielectrics discussed by Fok[3]. In this form, the problem was first posed
by Todes and Kontorova[4]; however, owing to their search
for an extremely general solution, their formulas not only
are not usable for numerical calculations but do not even
permit quantitative conclusions. The only concrete conclusion arrived at in that work was a correct relation between the critical pressure of ignition and the diameter
of the vessel which, however, can be obtained in a simpler
way, without analytical solution, from dimensional considerations.

We solve the problem on the following assumptions[5]:

(1)   We consider the temperature rise, prior
      to explosion, small compared with the
      absolute temperature of the walls,

$$\frac{\Delta T}{T} \ll 1;$$

(2)   The heat conductivity of the wall will
      be considered infinitely great;

(3)   The rate of the reaction will be con-
      sidered dependent on the temperature
      only through

$$e^{-\frac{E}{RT}}$$

      i.e., we shall disregard the temperature
      dependence of the pre-exponental factor,

the variation of the density in different
parts of the vessel, etc.

This third assumption can always be made without sacrifice
to the accuracy of the result. The first assumption is
equivalent, as will be shown later, to the condition
$RT \ll E$, and, therefore, it puts a definite limit to the
applicability of the theory. For reactions with auth-
entically homogeneous kinetics, it is practically almost
always legitimate in the whole range of conditions where
explosions can be observed. One can, however, point to
processes, entirely analogous to thermal explosion and
describable by the same equation (only, probably, not
with an Arrhenian temperature law of the rate) for which
assumption (1) is not legitimate - for example, the
spontaneous inflammation of a heap of coal. Clearly, our
theory is inapplicable to this kind of process with a
large pre-explosion temperature rise and long induction
periods. But our assumption is doubtlessly legitimate for
all instances which one usually has in mind when speaking
of thermal explosion.

With regard to assumption (2) one can, of course,
create conditions under which it will be incorrect - for
example, if one carries out explosions in capillaries with
thick walls made of a poor heat-conducting material. However, as the heat capacity of the wall is much greater
than that of the gas, even the total amount of heat evolved
in the explosion will raise the temperature of the walls
only insignificantly.

It would thus be entirely wrong to assume that
a stationary temperature distribution will establish
itself within the wall, as was done in the work of Todes
and Kontorova referred to above. Therefore, it is practically reasonable to assume in all cases that the initial
explosion temperature T_o is given at the inner surface
of the walls.

We shall first consider the case when convection
is entirely absent. The equation of heat conductance in
the stationary case and for a field with continuously distributed sources of heat of density Qv (where Q is the
thermal effect, and v the rate of the reaction) is of
the form

$$a\Delta T = - \frac{Q}{c\rho} v \qquad (VII, 1)$$

where a is the temperature conductivity of the gaseous
mixture, c its heat capacity, and ρ its density; Δ
is the Laplace operator. According to assumption (3) we
take the rate of reaction to be

$$v = ze^{-\frac{E}{RT}}$$

where E is the activation energy; bearing in mind that
$c\rho a = \lambda$, the heat conductivity of the gaseous mixture,
we can put equation (VII, 1) in the form

$$\Delta T = - \frac{Q}{\lambda} ze^{-\frac{E}{RT}} \qquad (VII, 2)$$

This equation has to be solved under boundary conditions
involving a constant temperature T_0 at the walls of the
vessel; by assumption (2), we can place this temperature
at the inner surface of the wall.

Solution of equation (VII, 2) satisfying the
boundary conditions, gives the stationary temperature distribution in the reaction vessel at the wall temperature
T_0. At a certain temperature, such a distribution becomes
impossible; this temperature we shall consider as the
ignition temperature. Its relation with the heat effect
and the rate of the reaction, the heat conductivity of the
mixture, and with the shape and dimensions of the vessel
can be found by an analysis of the properties of equation

(VII, 2) and of its solutions. Our problem is to find this relation.

As has been shown in the foregoing chapter with the aid of the similitude theory, the stationary temperature distribution must be of the form (VI, 15), i.e., must contain one dimensionless parameter δ, defined by formula (VI, 14). The task now is to find a concrete analytical form of this dependence. For an infinite vessel with plane parallel walls, equation (VII, 1) can be integrated in a general form for any temperature dependence law of the rate of reaction v. In this case, it takes the form

$$\frac{d^2T}{dx^2} = -\frac{Q}{\lambda} v(T)$$ (VII, 3)

The general integral of this equation for any forms of the function $v(T)$ involves two quadratures and is of the form

$$x = \int \frac{dT}{\sqrt{-2 \int \frac{Q}{\lambda} v(T) \, dT}}$$

with two arbitrary constants. The critical condition of ignition will be the totality of values of the parameters with which this expression cannot satisfy the boundary conditions at any value of the arbitrary constants. If one places the origin of the coordinates in the center of the vessel and designates the width of the latter by 2r, the boundary conditions will be formulated thus: at $x = \pm r$, $T = T_0$. Because of the symmetry, one can solve the equation for one-half of the vessel, under the combined boundary conditions, at $x = r$, $T = T_0$, and at $x = 0$, $\frac{dT}{dx} = 0$. In order to find the criterion by which the boundary conditions are satisfied, the temperature in the center of the vessel T_m is given as a parameter and the equation is solved under the Cauchy conditions, at

$x = 0$, $T = T_m$ and $\frac{dT}{dx} = 0$; the solution will then take the form

$$x = \int_{T_o}^{T_m} \frac{dT}{\sqrt{2 \int_T^{T_m} \frac{Q}{\lambda} v(T) \, dT}} \qquad \text{(VII, 4)}$$

It contains one variable parameter T_m determinable from the boundary condition

$$r = \int_{T_o}^{T_m} \frac{dT}{\sqrt{2 \int_T^{T_m} \frac{Q}{\lambda} v(T) \, dT}} \qquad \text{(VII, 5)}$$

There is no second arbitrary constant in (VII, 4) because the form of the solution (VII, 4) automatically satisfies the boundary condition, at $x = 0$, $\frac{dT}{dx} = 0$.

Let us designate the integral standing in the right-hand member of (VII, 5) by $\psi(T_m, T_o)$. If this integral is a monotonous function of T_m, the stationary state always is possible. If the form of the function $v(T)$ is such that, with changing T_m, ψ passes through an extreme, this extremum should give the critical condition of ignition. It gives directly the critical dimension of the vessel; at dimensions beyond the extreme condition (VII, 5) cannot be satisfied at any value of T_m. Physically it is obvious that the critical diameter of the vessel must be maximum, and that the extreme here discussed is a maximum.

The most general form of the critical condition of ignition for a plane-parallel vessel is thus

$$\left(\frac{\partial \psi}{\partial T_m} \right)_{T_o} = 0 \qquad \text{(VII, 6)}$$

At values of r greater than critical, a stationary
temperature distribution in the vessel is impossible. At
values of r less than critical, it is geometrically obvious that to each value of r there must correspond at
least two values of T_m, i.e., two different stationary
temperature distributions in the vessel. Its counterparts
in the elementary theory of Semenov[3] are two intersections
of the straight line of the heat removal and the curve of
the rate of reaction, degenerating at the limit into a
point of contact. From the analogy with Semenov's theory
we conclude that, of the two possible stationary temperature distributions, only that which corresponds to a
smaller T_m can be stable.

A stationary distribution is obtained formally
for any value of the temperature in the center of the
vessel. But not all these distributions are stable.

We now turn to the real form of the temperature
dependence of the rate of reaction. Introduction of
Arrhenius' law into equations (VII, 3) - (VII, 6) leads to
an expression which is not integrable in elementary functions and is very unsuitable for calculations*. But, on
the basis of assumption (1), we can pass from equation
(VII, 2) to (VI, 11) or (VI, 13), and the result will remain correct for small values of the parameter

$$\frac{RT_o}{E}$$

After that, equation (VII, 3) will take the form

$$\frac{d^2\vartheta}{dx^2} = -\frac{Q}{\lambda} ze^{-\frac{E}{RT_o}} e^{\frac{E}{RT_o^2}\vartheta} \qquad (VII, 7)$$

---

* Moreover, Arrhenius' law gives a third stationary
temperature distribution with no physical meaning, possible
at any r and corresponding to $T_m \sim \frac{E}{R}$.

or, in the dimensionless variables $\theta = \dfrac{E}{RT_o^2} \vartheta$ and

$\xi = \dfrac{x}{r}$, where r is the half-width of the vessel,

$$\frac{d^2\theta}{d\xi^2} = - \delta e^\theta \qquad\qquad (VII, 8)$$

where the parameter δ has the value

$$\delta = \frac{E}{RT_o^2} \frac{Q}{\lambda} \, r^2 z e^{-\dfrac{E}{RT_o}} \qquad\qquad (VII, 9)$$

The general integral of equation (VII, 8) has the form

$$e^\theta = \frac{a}{\cosh^2 \left(b \pm \sqrt{\dfrac{a\delta}{2}} \cdot \xi \right)} \qquad\qquad (VII, 10)$$

with two arbitrary constants a and b. By the symmetry condition

$$\left(\frac{d\theta}{d\xi}\right)_{\xi = 0} = 0 \qquad \text{or} \quad \theta(\xi) = \theta(-\xi)$$

the constant b must be equal to zero (this is the condition of equality of the temperatures of the two walls
of the vessel), and the integral becomes

$$e^\theta = \frac{a}{\cosh^2 \sqrt{\dfrac{a\delta}{2}}\,\xi} \qquad\qquad (VII, 11)$$

The arbitrary constant a is determined from the boundary
condition, at $\xi = 1$, $\theta = 0$, whence we get the transcendent equation

$$a = \cosh^2 \sqrt{\dfrac{a\delta}{2}} \qquad\qquad (VII, 12)$$

At values of δ for which (VII, 12) has a solution, a
stationary temperature distribution is possible, and its
form is found by substituting this solution in (VII, 11).
At values of δ for which (VII, 12) has no solution, explosion will take place. The critical condition of ignition will be determined by the value of δ at which
(VII, 12) ceases to have a solution. For the following it
is convenient to introduce, instead of the integration
constant a, a new magnitude σ, defined by the relation

$$a = \cosh^2 \sigma$$

The transcendent equation (VII, 12) then takes the form

$$\frac{\cosh \sigma}{\sigma} = (\tfrac{\delta}{2})^{-\frac{1}{2}} \qquad\qquad \text{(VII, 13)}$$

It is now easily seen that the critical condition of ignition is determined by the minimum value of the magnitude
$\frac{\cosh\sigma}{\sigma}$; this minimum lies at $\sigma_{cr} = 1.2$ and gives

$$\delta_{cr} = 0.88 \qquad\qquad \text{(VII, 14)}$$

as the critical condition of ignition.

The maximum pre-explosion temperature rise is
found from (VII, 11), with $\xi = 0$:

$$\theta_m = \ln a_{cr} = \ln \cosh^2 \sigma_{cr} = 1.2 \qquad\qquad \text{(VII, 15)}$$

$$(\Delta T)_m = 1.2 \frac{RT_o^2}{E} \qquad\qquad \text{(VII, 16)}$$

Thus, the problem of self-ignition for a plane-parallel
vessel is solved completely.

The analytical expression (VII, 11) enables us to
discuss also the stationary temperature distribution in the
course of the reaction in the vessel below the explosion

limit. With the value of the dimensionless temperature θ in the center of the vessel designated by θ_0, equation (VII, 11) takes the form

$$\theta = \theta_0 - 2 \ln \cosh \sigma \xi \qquad (VII, 17)$$

where

$$\theta_0 = 2 \ln \cosh \sigma \qquad (VII, 18)$$

The magnitude σ is a function of the parameter δ which expresses the totality of all the properties of the system (rate and heat of the reaction, heat conductivity, dimensions of the wall) and is determined from the transcendent equation (VII, 13). The solution of this equation is represented in Figure 19, where only the part of the curve up to the maximum corresponds to stable temperature distributions. At a value of $\sigma = 1.2$, corresponding to maximum δ, explosion occurs. Smaller values of σ correspond to smaller T_m and consequently, according to the above, correspond to stationary, and greater σ to unstable temperature distributions. If all the properties of the mixture and the conditions of the experiment are known, including the dimensions of the vessel, we can calculate the value of the parameter δ. If this value turns out to be less than 0.88, we find the

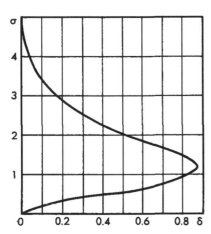

FIGURE 19. SOLUTION OF
THE TRANSCENDENT EQUATION
(VII, 13).
Ordinates are σ
Abscissas are δ

corresponding value of σ (the smaller of the two possible ones) from the curve in Figure 19. With it we can determine the temperature rise in the center of the vessel by equation (VII, 18), and the temperature distribution over the vessel by equation (VII, 17).

It now remains to find the critical value of δ and the value of the pre-explosion temperature rise for vessels of spherical and of cylindrical shape. For these purposes, equation (VI, 13) takes the form

$$\frac{d^2\theta}{d\xi^2} + \frac{1}{\xi}\frac{d\theta}{d\xi} = -\delta\, e^\theta \qquad\qquad (VII,\ 19)$$

for an infinitely long cylindrical vessel, and

$$\frac{d^2\theta}{d\xi^2} + \frac{2}{\xi}\frac{d\theta}{d\xi} = -\delta\, e^\theta \qquad\qquad (VII,\ 20)$$

for a spherical vessel. Here $\xi = \frac{x}{r}$, where r is the radius of the vessel. The boundary conditions are, at $\xi = 1$, $\theta = 0$, and at $\xi = 0$, $\frac{d\theta}{d\xi} = 0$.

Equations (VII, 19) and (VII, 20) cannot be integrated in terms of elementary functions. A general integral would be, by the way, meaningless in principle, as it has a singular point at $\xi = 0$, i.e., in the center of the vessel, whereas we are interested only in solutions giving a finite θ at $\xi = 0$.

To find an approximate critical value of δ , we shall make use of numerical integration of equations (VII, 19) and (VII, 20), with the temperature θ_0 in the center of the vessel given as a parameter. In this, it is convenient to introduce the new variables

$$x = \xi\, \sqrt{\delta\, e^{\theta_0}}$$

$$y = \theta_0 - \theta$$

The integration is carried out under the boundary cond-
ditions at $x = 0$, $y = y' = 0$. To each pair of values
of x and y, satisfying the equation (VII, 19) or
(VII, 20), correspond the values

$$\delta = x^2 e^{-y}$$

$$\theta_o = y$$

In this way, the task is to find θ_o as a function of
δ. The variables x and y have an auxiliary significance.

It is convenient to begin the integration with
the aid of the series

$$y = \frac{1}{4} x^2 - \frac{1}{64} x^4 + \frac{1}{768} x^6 - \cdots$$

for a cylindrical vessel, and

$$y = \frac{1}{6} x^2 - \frac{1}{120} x^4 + \frac{1}{1890} x^6 - \cdots$$

for a spherical vessel. Further integration was done
numerically by the method of Adams. The results of the
numerical integration are given in Table 4 for a cylindrical and in Table 5 for a spherical vessel. The tables
give the relation between δ and θ_o.

As in the above discussed case of a plane-
parallel vessel, δ as a function of θ_o has a maximum.
Two stationary temperature distributions correspond to
each value of δ. That which corresponds to the smaller
value of θ_o is stable. The maximum value of δ gives
the critical condition of ignition, and the corresponding
value of θ_o, the maximum pre-explosion temperature rise.
The dependence of θ_o on δ gives the stationary temperature rise below the explosion limit.

From Tables 4 and 5 we find the critical condition

TABLE 4

Cylindrical Vessel

δ	θ_0	δ	θ_0	δ	θ_0	δ	θ_0
0.0000	0.0000	0.7909	0.2346	1.7792	0.8102	1.9944	1.5070
0.0100	0.0025	0.9137	0.2809	1.8341	0.8774	1.9845	1.5775
0.0396	0.0100	1.0354	0.3299	1.8803	0.9455	1.9712	1.6477
0.0880	0.0224	1.1531	0.3823	1.9176	1.0146	1.9548	1.7178
0.1538	0.0396	1.2658	0.4372	1.9486	1.0839	1.9351	1.7875
0.2351	0.0615	1.3718	0.4946	1.9706	1.1541	1.9135	1.8565
0.3297	0.0879	1.4705	0.5545	1.9874	1.2243	1.8896	1.9254
0.4361	0.1188	1.5612	0.6158	1.9967	1.2949	1.8646	1.9939
0.5487	0.1538	1.6423	0.6794	2.0008	1.3657	1.8368	2.0618
0.6685	0.1920	1.7155	0.7441	1.9999	1.4365	1.8075	2.1293

TABLE 5

Spherical Vessel

δ	θ_0	δ	θ_0	δ	θ_0	δ	θ_0
0.0000	0.0000	1.4547	0.2981	3.0027	0.9598	3.3188	1.6702
0.0100	0.0017	1.6062	0.3382	3.0587	1.0114	3.3153	1.7186
0.0397	0.0067	1.7513	0.3797	3.1086	1.0631	3.3086	1.7664
0.0887	0.0150	1.8941	0.4225	3.1521	1.1148	3.3007	1.8138
0.1558	0.0265	2.0308	0.4672	3.1887	1.1665	3.2904	1.8611
0.2399	0.0411	2.1620	0.5127	3.2213	1.2180	3.2782	1.9075
0.3394	0.0590	2.2856	0.5596	3.2484	1.4695	3.2671	1.9536
0.4525	0.0796	2.4021	0.6075	3.2707	1.3206	3.2509	1.9992
0.5771	0.1035	2.5115	0.6561	3.2880	1.3716	3.2375	2.0443
0.7116	0.2295	2.6122	0.7056	3.3020	1.4222	3.2200	2.0889
0.8531	0.1589	2.7060	0.7555	3.3211	1.4726	3.2042	2.1330
1.0005	0.1901	2.7912	0.8052	3.3173	1.5227	3.1854	2.1766
1.1508	0.2241	2.8689	0.8571	3.3216	1.5722	3.1668	2.2197
1.3026	0.2603	2.9393	0.9082	3.3217	1.6214	3.1490	2.2623

of ignition

$$\delta_{cr} = 2.00$$

for a cylindrical vessel and by

$$\delta_{cr} = 3.32$$

for a spherical vessel. The maximum pre-explosion temperature rise in a cylindrical vessel amounts to

$$1.37 \frac{RT_o^2}{E}$$

and in a spherical vessel to

$$1.60 \frac{RT_o^2}{E}$$

We have found the final form of the critical condition of ignition for vessels of plane-parallel, cylindrical, and spherical shapes on the assumption of purely convective heat exchange. It comes down to a definite numerical value of the dimensionless parameter δ, equal to 0.88 for a plane-parallel, 2.00 for a cylindrical, and 3.32 for a spherical vessel.

Let us confront the results obtained with the theory[1,2] in which the temperature is assumed to be constant over the whole volume of the vessel and the heat flow proportional to some heat exchange coefficient α, assumed to be determinable experimentally. In this theory, one has for the ignition limit the expression see formula (VI, 21) in the preceding chaper

$$\frac{E}{RT_o^2} Qze^{-\frac{E}{RT_o}} = \frac{\alpha S}{\omega\theta} \qquad (VII, 21)$$

where S is the surface area of the vessel and ω its
volume. It is easily seen that this expression coincides
with ours except for the numerical factor or rather the
not yet accurately determined magnitude α. Thus, all
qualitative relationships, and the relations between the
different explosion parameters are the same in our theory
as in Semenov's; but we are enabled to calculate also the
absolute values of the critical parameters in the range
of conditions where the heat exchange can be considered
purely conductive. The concepts of Semenov's theory are
convenient in that they enable one to build up the nonstationary theory with allowance for the loss through combustion over the induction period and for the particularly
important self-acceleration of autocatalytic reactions.
In view of the mathematical complexity of the integration
of the complete nonstationary partial differential equation
with both the temperature, coordinate and time dependence
taken into account, one can propose the following approximate method of solution of nonstationary problems. From
a comparison of equations (VII, 21) and (VII, 9) and our
critical value of δ, we will find the effective values
of the heat exchange coefficient; with these substituted
in the elementary theory, it will give for the stationary
case results identical with the exact theory. In the
following, we shall use these effective heat exchange coefficients in the investigation of nonstationary cases.

The values of the effective heat exchange coefficients found in this way are

$$\alpha_{eff} = 4.8 \frac{\lambda}{d}$$

for a plane-parallel vessel,

$$\alpha_{eff} = 5.4 \frac{\lambda}{d}$$

for a cylindrical vessel, and

$$\alpha_{eff} = 5.9 \frac{\lambda}{d}$$

for a spherical vessel where d is the diameter of the
vessel and λ the heat conductivity of the gaseous mixture.

One can proceed in an analogous way also to find
the critical condition of ignition for systems of a complex geometric shape, for example, a cylinder of finite
length for which an integration of equation (VI, 13) would
be difficult. For this purpose we can compare the quasistationary values of the characteristic time of cooling,
obtained by integration of the usual nonstationary equation
of heat conductance, with the values which, upon substitution in (VII, 21), make it coincide exactly with the
correct condition of ignition.

Combining formula (VII, 21) with the definition
of the parameter δ (VI, 14), we get

$$\delta = \frac{1}{4e} \frac{\alpha_{eff} \, d^2}{\lambda} \frac{S}{\omega} \qquad (VII, 22)$$

In order to find approximate values of δ for vessels of
any given geometric shape, we can use instead of α_{eff} the
heat exchange coefficient $\bar{\alpha}$ referred to the mean temperature difference over the whole volume.

According to (VI, 18) one has, in the absence of
evolution of heat

$$\frac{d\bar{\vartheta}}{dt} = - \frac{\bar{\alpha}S}{c\rho\omega} \, \bar{\vartheta} \qquad (VII, 23)$$

where $\bar{\vartheta}$ is the mean temperature difference over the
whole volume. Having found the magnitude $\frac{d\bar{\vartheta}}{dt}$ from the
quasistationary solution of Fourier's heat conductance
equation, we can determine the magnitude α and, substituting it for α_{eff} in (VII, 22), find the approximate
value of δ which we shall designate by δ^*.

Let us test this procedure in the instance of a spherical vessel. The Fourier equation of heat conductance in this instance is of the form

$$\frac{\partial \vartheta}{\partial t} = \frac{a}{r} \frac{\partial^2}{\partial r^2} (r\vartheta)$$

where $a = \frac{\lambda}{c\rho}$ is the temperature conductivity, and $\vartheta = T - T_0$.

The solution which remains finite at $r = 0$, will be

$$\vartheta = \sum_k A_k e^{-k^2 at} \frac{\sin kr}{r}$$

The values of k should satisfy the boundary condition, at $r = \frac{d}{2}$, $\vartheta = 0$; this gives

$$k\frac{d}{2} = n\pi; \qquad k = \frac{2n\pi}{d}$$

where n is an integer.

A quasistationary state of heat exchange will be established when the higher terms of the series can be disregarded. Then

$$\vartheta \approx A_1 e^{-\frac{4\pi^2 at}{d^2}} \frac{\sin \frac{2\pi}{d} r}{r}$$

Differentiation gives

$$\frac{d\vartheta}{dt} = -\frac{4\pi^2}{d^2} a\vartheta$$

Combining with (VII, 23), we get

$$\overline{\alpha} = \frac{4\pi^2}{d^2} \frac{\omega}{\mathfrak{S}} \lambda$$

$$\frac{\bar{\alpha} d^2}{\lambda} \frac{S}{\omega} = 4\pi^2$$

and, introducing in (VII, 22)

$$\delta^* = \frac{\pi^2}{e} = 3.64$$

In the instance of an infinite cylindrical vessel the solution of the nonstationary equation will be

$$\vartheta = \sum_k A_k e^{-k^2 at} I_0(kr)$$

where I_0 is the Bessel function of zero order.

The boundary conditions give

$$k = \frac{2\mu_k}{d}$$

where μ_k are the roots of the Bessel function of zero order.

The quasistationary solution is of the form

$$\vartheta = A_1 e^{-\frac{4\mu_1^2 at}{d^2}} I_0\left(\frac{2\mu_1}{d} r\right)$$

where $\mu_1 = 2.4048$.

Further calculation gives

$$\frac{d\vartheta}{dt} = -\frac{4\mu_1^2 a}{d^2} \vartheta$$

$$\frac{\bar{\alpha} S}{c\rho\omega} = \frac{4\mu_1^2}{d^2} a$$

$$\frac{\alpha d^2}{\lambda} \frac{S}{\omega} = 4\mu_1^2$$

The corresponding value of δ will be

$$\delta* = \frac{\mu_1^2}{e} = 2.14$$

The ratio $\frac{\delta*}{\delta}$ equals 1.09 for a spherical and 1.07 for a cylindrical vessel. On that basis, we can put for vessels of any given shape

$$\delta = \frac{\delta*}{1.08} = \frac{1}{1.08}\frac{1}{4e}\frac{\bar{a}d^2}{\lambda}\frac{S}{\omega} \qquad\qquad (VII, 24)$$

In the case of a cylindrical vessel of finite length L, the solution of the Fourier equation is

$$\vartheta = \underset{1}{\Sigma}\ \underset{m}{\Sigma}\ A_{ml}e^{-(1^2 + m^2)at}\cos 1z\ I_o(mr)$$

and the boundary conditions give

$$1\frac{L}{2} = (n + \frac{1}{2})\pi;\qquad \frac{md}{2} = \mu_m$$

The quasistationary solution will be

$$\vartheta = A_{11}e^{-\left(\frac{\pi^2}{L^2} + \frac{4\mu_1^2}{d^2}\right)at}\cos\frac{\pi z}{L}\ I_o\ (\frac{2\mu}{d}\ r)$$

Differentiation gives

$$\frac{d\vartheta}{dt} = -\left(\frac{\pi^2}{L^2} + \frac{4\mu_1^2}{d^2}\right)a\vartheta$$

$$\frac{\bar{a}S}{c\rho\omega} = \frac{\pi^2}{L^2} + \frac{4\mu_1^2}{d^2}$$

$$\delta_{cr} = \frac{1}{1.08}\cdot\frac{1}{4e}\left(4\mu_1^2 + \pi^2\frac{d^2}{L^2}\right) = 2.00 + 0.843\ (\frac{d}{L})^2 \quad (VII, 25)$$

We shall make use of formula (VII, 25) in the evaluation
of experimental results of ignition in cylindrical vessels
with a small ratio of length to diameter. At a length
equal to twice the diameter, the correction is no more
than 10 percent, and at $\frac{L}{d} = 6$ it is about 1 percent.

In the foregoing discussion it was assumed that
the heat transfer takes place by way of pure conduction,
with convection entirely absent. This assumption is
justified only in the limiting case of small vessel dimensions and low initial pressures. To what extent concrete experimental conditions approximate the limiting
case, can be decided only by way of comparison of the
theory with experimental data.

Correction for the Loss Through Combustion over the Induction Period

We have formulated the condition of ignition as
the condition under which a stationary temperature distribution in the reacting mixture becomes impossible. The
concentration of the reactants was taken equal to the initial concentration. In reality, explosion occurs after
the lapse of a certain induction period during which the
reactant concentration has had time to change. The resulting shift of the ignition limit has been calculated by
us recently[17]. We showed that the change of the relative
rate of reaction at the ignition limit, as a result of
loss of reactant through combustion over the induction
period, is expressed by

$$\epsilon = \sqrt[3]{\frac{2\pi^2 m^2}{e^2 B^2}} = 1.39 \left(\frac{m}{B}\right)^{2/3} \qquad \text{(VII, 26)}$$

where m is the order of the reaction, and B the dimensionless maximum temperature of the explosion, defined
by formula (VI, 25).

The ignition limit corrected for the loss through

combustion can be found with the aid of the corrected value of the above-introduced parameter δ

$$\delta' = \delta_0 (1 + \epsilon) \qquad\qquad (VII, 27)$$

The ignition temperature at constant pressure and composition of the mixture will vary by

$$\Delta T = \frac{RT_0^2}{E} \epsilon \qquad\qquad (VII, 28)$$

Thermal Explosion in Autocatalytic Reactions

In autocatalytic reactions, it is necessary to
distinguish between the thermal and the chemical (autocatalytic) induction period. By thermal induction period
we mean, here and always, the time during which heat is
accumulated and the temperature rises in the reacting mixture. The chemical induction period is the time during
which active products accumulate until the maximum rate of
reaction is reached.

Correspondingly, in autocatalytic reactions, the
position of the ignition limit can be considered as a
function of the induction period. The longer is the accepted induction period, the more active products will have had
time to accumulate, the greater will be the rate of reaction, and the easier the ignition. Thus, if we prescribe
some arbitrarily chosen value of the induction period, we
get the corresponding curve of the ignition limit for that
induction period; it will lie the lower, the longer is the
induction period.

However, with increasing length of the induction
period, we shall reach a maximum value of the rate of reaction; on further lengthening time of reaction, its rate
will fall owing to consumption of the reactant. In the
simplest case of an autocatalysis of the first order, both

with respect to the initial reactant and the final product,
the maximum rate of reaction will be attained after consumption of half the initial amount of reactant.

The absolute (lowest) ignition limit for an autocatalytic reaction is found by substituting this maximum
value of the rate of reaction in the above-derived condition of ignition. This limit corresponds to the lowest
values of temperature and pressure below which ignition is
impossible at any length of the induction period. At the
lowest limit, the induction period will have a maximum
length, corresponding to the time of attainment of the
maximum rate of reaction. In the calculation of the absolute (lowest) limit of ignition for autocatalytic reactions, the correction for loss through combustion, introduced in the foregoing paragraph, is superfluous, since
the change of reactant concentration has been taken into
account very exactly in the calculation of the maximum
rate of reaction.

Experimental Verification of the Theory
of Thermal Explosion

The theory just discussed permits advance calculation of the position of the ignition limit for reactions with known kinetics. This possibility was put to
use, both by ourselves and other authors in the USSR and
abroad, in the solution of a variety of problems. First
of all, we have[5] tested the theory against some reactions
for which the kinetics, the ignition limit, and other data
necessary for the calculation were well known from the
literature and for which there was every reason to assume
beforehand that the explosion is of thermal nature. In
this category fall the reactions, studied by Rice[6], of
decomposition of azomethane and of ethyl azide; the decomposition of methyl nitrate (Apin and Khariton[7]), and
the oxidation of hydrogen sulfide (Yakovlev). The first
three reactions follow simple monomolecular, the fourth
relatively complex autocatalytic kinetics. In three

cases, calculation gave results in very good agreement
with the experiment; there were some discrepancies in
the case of ethyl azide, the reasons for which are not
clear.

The theory was further applied to the predic-
tion of the position of the ignition limit where it had
not been previously observed experimentally. The previously completely unknown thermal ignition of nitrous
oxide could be successfully predicted. The experimental
ignition limit, determined by Zeldovich and Yakovlev[8],
agreed very well with the theoretically precalculated limit.

The results of the comparison of the theory with
the experimental data for these reactions are given in
Tables 6, 7, 8 and 9. It is advisable to compare the observed and the calculated ignition temperature since
small errors in the temperature produce a considerable
change in these magnitudes - because of the exponential
temperature dependence of δ and the critical pressure.

TABLE 6

Decomposition of Azomethane (Rice)

$$(CH_3)_2 N_2 \longrightarrow C_2H_6 + N_2$$

p(mm)	$T^{\circ}K$ Calc.	$T^{\circ}K$ Obs.
191	619	614
102	629	620
67	635	626.3
55	638	630.7
38	644	636.4
31	647	643.4
28	649	644.9
22.5	653	651.2
18	656	659

TABLE 7

Decomposition of Methyl Nitrate
(Apin and Khariton)

$$2CH_3ONO_2 \longrightarrow CH_3OH + CH_2O + 2NO_2$$

p(mm)	$T^O K$ Calc.	$T^O K$ Obs.
4.2	590	597
5.5	586,	584
8.5	578	567
12.5	572	553
16.5	566	546
33.5	556	534
45.4	551	529
87.0	541	522.5
107.0	538	521
163.0	531	519.5

TABLE 8

Oxidation of H_2S (Yakovlev)

$$2H_2S + 3O_2 \longrightarrow 2H_2O + 2SO_2$$

p(total) mm	p(H_2S) mm	T Calc.	T Obs.
244	98	544	578
400	160	523	552
745	298	499	525

TABLE 9

Decomposition of N_2O (Zeldovich and Yakovlev)

$$2N_2O \longrightarrow 2N_2 + O_2$$

p(mm)	T Calc.	T Obs.
170	1255	1285
330	1175	1195
590	1110	1100

To calculate the ignition temperature from given pressure and dimensions of the vessel it is necessary to solve the transcendent equation

$$\frac{Q}{\lambda} \frac{E}{RT^2} r^2 z e^{-\frac{E}{RT}} = \delta_{cr}$$

Inasmuch as the temperature which stands before the exponential has only an insignificant effect on the value of the left-hand member as compared with the exponential factor, it is easy to solve this equation in the logarithmic form by the method of successive approximations.

The values of the thermal and kinetic constants were taken as far as possible from the kinetic work referred to. For the heat conductivities of azomethane, ethyl azide, and methyl nitrate, we took 1.0×10^{-4} cal degree^{-1} sec^{-1} cm^{-1}. In contrast to the other four reactions, experimental data for ethyl azide yield values of δ of about 20, as against the theoretical value of 3.32. In the face of such a discrepancy with the theory, there is no point in comparing the calculated and the observed ignition temperature for this reaction.

The reactions of decomposition of azomethane, methyl nitrate, ethyl azide, and nitrous oxide, follow simple monomolecular kinetics. Calculations for these

reactions were made by the initial rates of reaction,
without allowing for the loss through combustion over the
induction period. Of the legitimacy of this simplification, it is easy to satisfy oneself with the aid of formulas (VII, 26) - (VII, 28). For the decomposition of azomethane, according to Rice[6], Q = 43000 cal/mole,
E = 51200 cal/mole, c_v = 25.7 cal/degree·mole; the explosion
was studied in the temperature range 614-659°K. Hence, at
the mean temperature of 636°K, we find the value of the
parameter B = 106, and formula (VII, 26) gives
ϵ = 6.2 × 10^{-2}. The ignition temperature, calculated by
the stationary theory, will vary by

$$\Delta T = \frac{RT_o^{\,2}}{E} \, \epsilon = 0.97°C$$

For ethyl azide, according to the same author, Q = 55000,
E = 39000, c_v = 25.3, temperature range 533-630°K.
Hence, at the mean temperature of 548°K, we get the values
B = 141, ϵ = 5.1 × 10^{-2}, and ΔT = 0.78°C.

For the decomposition of nitrous oxide
Q = 19500, E = 53000, temperature range 1100-1285°K, mean
temperature 1192°K; the heat capacity at that temperature
can be estimated with the aid of the formula of Planck-
Einstein. The N_2O molecule is linear. Its frequencies
are: 589 cm^{-1} (doubly degenerate), 1285 cm^{-1}, and
2224 cm^{-1}. The heat capacity c_v at 1192°K is

$$c_v = \frac{5}{2} R + 2\Phi(0.705) + \Phi(1.541) + \Phi(2.67)$$

$$= 11.56 \text{ cal/mole°C}$$

(Φ is the Planck-Einstein function). Hence we get
B = 32, ϵ = 0.14, ΔT = 7.4°C. Only for this reaction does
the correction for the loss over the induction period exceed the limits of experimental error. This correction
would somewhat improve the agreement between theory and

experiment, as can be seen from Table 9.

The reaction of oxidation of hydrogen sulfide
is outspokenly autocatalytic. For this reaction calculations were done by the maximum rate of reaction.

An equally successful result was obtained also
for the so-called third ignition limit of mixtures of
oxygen with hydrogen. With regard to this limit, the most
contradictory statements have been made in the literature.
In particular, some authors are of the opinion that the
third limit is of a chain nature. Owing to some experimental difficulties, the very position of that limit was
not known with accuracy until recently. With the aid of
the data of Chirkov[9] on the kinetics of the reaction below the limit, we have calculated the position of the
third limit on the assumption that it is of a purely
thermal nature. The reaction is autocatalytic; the calculation was done by the maximum rate of reaction. For
quite a long time, no data were available for a comparison
with experiments. From the experiments of Chirkov it
appeared that, undoubtedly, no explosion occurs below the
limit. Subsequently, Oldenberg and Sommers[10] have published results which show that explosion decidedly does
occur somewhat above the limit which we had calculated.
Chirkov's experiments were carried out in untreated vessels
of "durabax" glass of 5 cm diameter. Our calculations refer to these experimental conditions.

In the work of Oldenberg and Sommers the conditions were different: the experiments were carried out
in vessels treated with a solution of potassium chloride.
As was recently remarked by Voevodskii, the reaction
velocity in such vessels is very much slower than in untreated vessels . Consequently, the limit of thermal ignition can be strongly shifted to higher temperatures and
pressures, and it is doubtful whether it is legitimate to
confront the observations of Oldenberg and Sommers with
our calculations, based on the kinetics measured in

untreated vessels. Only recently, Ziskin[11] has determined the exact position of the lowest ignition limit under atmospheric pressure. He worked with a stream, in tubes of different diameters, i.e., under conditions under which the previously mentioned experimental difficulties are eliminated. The experimentally observed ignition limit coincided, with a satisfactory accuracy, with the limit which we had calculated. In an earlier publication of Pease[12], there is also an indication of an ignition limit which agrees perfectly with our calculation.

Figure 20, taken from the article of Ziskin, represents the dependence of the ignition temperature under atmospheric pressure on the diameter of the vessel. The full line was calculated by our formula. The circles represent the experimental data of Ziskin; the straight cross (+) corresponds to the result of Pease, and the oblique cross (x) to the point calculated for a vessel of 5 cm diameter, taken from our work. The exceptionally good agreement between the experimental data and the calculations by the stationary theory is apparently

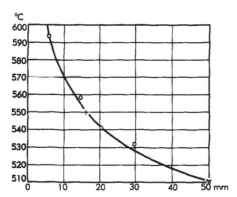

FIGURE 20.   IGNITION LIMIT OF THE DETONATING MIXTURE, AFTER ZISKIN

Ordinates are ignition temperature in °C.
Abscissas are vessel diameters in mm.

The full line was drawn by our theoretical formula (VI, 16). The circles represent the experimental results of Ziskin, and the cross + the result of Pease.

due to the autocatalytic character of the process. For autocatalytic reactions, good agreement between theory and experiment can be expected, thanks to the absence of a correction for a loss by combustion. Explosion occurs after the maximum reaction velocity has been attained and the mixture has had time to heat up.

Kokochashvili in his thesis has calculated, and then determined experimentally, the ignition limit for mixture of hydrogen and bromine from the long and well known kinetics of that reaction. Besides the agreement between theory and experiment, the thermal interpretation of the dependence of the ignition limit on the composition, due to the change of the heat conductivity, is of interest.

Harris[13] has applied our theory to interpret the nature of the explosion of ethyl peroxide. This problem is quite essential for the theory of the oxidation of hydrocarbons. Formerly[14] it was assumed that the explosion of the peroxide is a chain process, and that the large amounts of radicals formed in this explosion can bring about the ignition of the hydrocarbon; the formation of a cool flame was interpreted as an ignition of the peroxide accumulated to a critical concentration sufficient for the explosion. Harris has convincingly demonstrated the inconsistency of these ideas by showing that the condition of ignition of the peroxide is in close agreement with that calculated by our method of the assumption of a purely thermal nature of the explosion.

Rice[6] tested the theory of the thermal ignition against the experiment by a different method, namely by the calculation of the length of the induction period. We have shown[16] Rice's method to involve a series of shortcomings and therefore not to be reliable on either theoretical or experimental grounds. As a result of a discussion, Rice[16] conceded the essence of our arguments and adopted our method of calculation, illustrating its good agreement with experimental data in a number of instances.

In our recent work with Blyumberg[18] we have determined the ignition limit in the thermal decomposition of acetylene. An analysis of the curves of change of pressure in the induction period showed that the process which leads to the explosion is the thermal dimerization of the acetylene, the kinetics of which we have studied

in detail, and that the ignition is of a thermal nature.
The position of the ignition limit could not be precalculated in this instance for lack of knowledge of heat
effect of the reaction. We solved the reverse problem,
namely, we calculated the heat effect of the dimerization
from the position of the ignition limit and the known
kinetics of the reaction. The calculation, without allowance for the loss through combustion during the induction
period, gave a heat effect of 64600 cal/mole dimer. With
the correction, which in this instance is quite significant, the corrected heat effect is 78500 cal/mole dimer.

Combining this result with the known rules of
the thermochemistry of organic compounds, we come to the
conclusion that the acetylene dimer, which we called
"polygen", must have a cyclic structure and its molecule
must contain two simple and two double bonds. The most
probable structure is that of cyclobutadiene. On the
basis of these results, we have proposed a "polygen
theory" which attributes to the polygen an essential role
in all pyrogenetic processes as a primary product of
polymerization and an initial substance for the formation
of resin and carbon. As a practical conclusion, we proposed the use of small additions of nitric oxide, as
inhibitor of the formation of polygen, to suppress the
resins and to increase the yield of unsaturated gases in
pyrolysis.

Literature

1. SEMENOV, Z. physik. Chem. 42, 571 (1928); Tsepnye
    reaktsii (Chain reactions) p. 116, Leningrad
    1934.

2. TODES, Zhur. Fiz. Khim. 4, 78 (1933).

3. FOK, Trudy Leningrad. Fiz Tekh. Lab. 5, 52 (1928).

4. TODES and KONTOROVA, Zhur. Fiz. Khim. 4, 81 (1933).

5. FRANK-KAMENETSKII, Zhur. Fiz. Khim. 13, 738 (1939).

6.   RICE, J. Am. Chem. Soc. 57, 310, 1044, 2212 (1935);
     J. Chem. Phys. 7, 701 (1939).

7.   APIN, TODES, KHARITON, Zhur. Fiz. Khim. 8, 866
     (1936).

8.   ZELDOVICH and YAKOVLEV, Doklady Akad. Nauk
     S.S.S.R. 19, 699 (1938).

9.   CHIRKOV, Acta physicochim. 6, 915 (1937).

10. OLDENBERG and SOMMERS, J. Chem. Phys. 7, 279 (1939).

11. ZISKIN, Doklady Akad. Nauk S.S.S.R. 34, 279 (1942).

12. PEASE, J. Am. Chem. Soc. 52, 5107 (1930).

13. HARRIS, Proc. Roy. Soc. A 175, 254 (1940).

14. NEIMAN, Uspekhi Khim. 7, 341 (1938).

15. FRANK-KAMENETSKII, J. Chem Phys. 8, 125 (1940).

16. RICE, J. Chem. Phys. 8, 727 (1940).

17. FRANK-KAMENETSKII, Zhur. Fiz. Khim. 20, 139 (1946).

18. BLYUMBERG and FRANK-KAMENETSKII, Zhur. Fiz. Khim.
     20, 1301 (1945).

CHAPTER VIII: FLAME PROPAGATION

If the combustion process has started at one point, it is able to spread over the space which contains the combustible mixture. The mechanism of this flame propagation can be of two kinds. There is the so-called normal, or still flame propagation, and detonation.

The mechanism of normal flame propagation involves transfer of heat conduction, or of active products through diffusion. In the detonation flame propagation, the combustion process is transmitted from one point to another not through direct transfer of heat or matter, but as a result of inflammation of the combustible mixture by a sudden density increase (shock wave).

The theory of detonation is closely bound with gas dynamics; we shall not consider it here. In contrast, the problem of normal flame propagation is a typical problem of the theory of combustion in the sense in which it is thought of in this book, i.e., it consists in simultaneous solution of equations of chemical kinetics and of heat transfer (or diffusion).

Equation and Boundary Conditions

We shall consider the problem in its generality, for both thermal and diffusional flame propagation, at

the same time since the two cases are entirely analagous
from the mathematical point of view.

We shall designate by x the main variable,
which is the temperature in the thermal, and the active-
product concentration in the diffusional flame propagation.
As zero point, we shall take the value of the corresponding magnitude in the initial mixture with the initial
temperature T_O or concentration C_O.

As natural yardstick, we shall take the maximum
temperature or concentration difference on completion of
the reaction. Thus, for thermal flame propagation,

$$x = \frac{T - T_O}{T_m - T_O} \qquad\qquad (VIII, 1)$$

and for diffusional flame propagation

$$x = \frac{C - C_O}{C_m - C_O} \qquad\qquad (VIII, 1a)$$

where T_m is the maximum adiabatic temperature of the
reaction, and C_m the maximum value of the active-product
concentration upon completion of the reaction.

The rate of reaction will be expressed as

$$v = \frac{Q(x)}{\tau_m} \qquad\qquad (VIII, 2)$$

where τ_m is the characteristic reaction time at the maximum temperature or active-product concentration, and
$Q(x)$ is a function of temperature or concentration, which
is given by the chemical kinetics.

We will, furthermore, introduce the dimensionless coordinate ξ, taking as the natural yardstick of
length the magnitude $\sqrt{a\tau_m}$ for thermal, and $\sqrt{D\tau_m}$ for
diffusional flame propagation.

With y designating the derivative of the variable x with respect to the dimensionless coordinate ξ, equation (VI, 31) or the corresponding equation for diffusional propagation will take the dimensionless form

$$y \frac{dy}{dx} - \mu y + Q(x) = 0 \qquad \text{(VIII, 3)}$$

where μ is the dimensionless flame propagation velocity, related with the true velocity w by

$$\mu = w \sqrt{\frac{\tau_m}{a}} \qquad \text{(VIII, 4)}$$

for thermal, and

$$\mu = w \sqrt{\frac{\tau_m}{D}} \qquad \text{(VIII, 5)}$$

for diffusional flame propagation.

The task of an analytical solution consists in finding values of μ for which the solution of equation (VIII, 3) satisfies the boundary condition

$$y = 0 \quad \text{at} \quad x = 0 \quad \text{and} \quad x = 1 \qquad \text{(VIII, 6)}$$

Uniqueness of the Solution

First of all it is necessary to ascertain whether such a solution is unique.

Many authors[1,2,3] sought, for different concrete forms of the function $Q(x)$, particular values of μ for which the solution of equation (VIII, 3) satisfies the conditions (VIII, 6). There was, however, no proof that the values of μ thus found were unique and, consequently, there could be no certainty that the value found for the velocity of the stationary flame propagation is physically obligatory.

That, in some cases, there can exist an infinite number (a continuum) of values of μ for which the solution of equation (VIII, 3) satisfies the conditions (VIII, 6), was first demonstrated mathematically by Kolmogorov, Petrovskii and Piskunov[4].

However, this infinite variety of solutions, formally satisfying the boundary conditions, actually has no physical meaning. An infinity of solutions is found only in cases when the function $Q(x)$ vanishes exactly at $x = 0$ and at the same time has a positive derivative. Physically, this corresponds to a situation where the initial state $(x = 0)$ is unstable even towards the slightest igniting impulse.

In order to have only one solution satisfying the boundary conditions, it suffices to assume that the rate of reaction, i.e., the function $Q(x)$, either vanishes at $x > 0$ or has a negative derivative at $x = 0$. This solution corresponds to the true velocity of flame propagation.

If the function $Q(x)$, i.e., the rate of reaction, does not vanish at $x = 0$, no stationary flame propagation is possible at all. In this case, chemical combustion reactions will take place in the initial mixture already at the initial temperature or active-product concentration and, consequently, the combustion process will sooner or later start at any point of the space regardless of whether or not the flame has reached it.

A stationary flame propagation can be obtained only when the kinetics of the reaction is prescribed in such a way that its rate in the initial mixture $(x = 0)$ is zero. Such a description of the real phenomenon can be either exact or approximate. In the most important case of thermal flame propagation, the rate of reaction depends on the temperature according to Arrhenius' law and, formally, does not become zero at the initial

temperature T_0. Practically, however, by reason of the
exponential nature of the Arrhenius law, the rate of reaction decreases so rapidly with falling temperature that
it can be considered negligibly small even at temperatures
distinctly higher than T_0. This immediately ensures the
uniqueness of the solution satisfying the boundary
conditions.

If the rate of reaction becomes zero exactly at
$x = 0$ and becomes at once positive at $x > 0$, we have a
case which lies at the limit between the possibility and
the impossibility of a stationary state. The reaction
cannot proceed in the initial mixture, but an infinitely
slight igniting impulse is sufficient to arouse it. Inasmuch as the introduction of an infinitely slight impulse
does not shake the legitimacy of the equation, we can con-
ceive of a "fictitious flame propagation" consisting in a
uniform artificial spread of such an infinitely faint
source of ignition at any velocity greater than the normal
flame propagation velocity. This will not alter the flame
propagation equation, and such a process will formally
satisfy both the equation and the boundary conditions.

Consequently, in this case, the equation and the
boundary conditions will be satisfied by an infinity of
values of μ, corresponding to any arbitrary values of
the flame propagation velocity greater than the true normal
velocity. This physical fact corresponds exactly to the
above discussed mathematical fact of existence of a continuum of values of μ, satisfying the equation and the
boundary conditions.

This holds only if the function $Q(x)$ vanishes
exactly at $x = 0$ and has at that point a positive derivative. If $Q(x)$ vanishes at $x > 0$ and has a negative
derivative at $x = 0$, the solution is unique. The
question of choice of a definite form of the function
$Q(x)$ is one of the choice of a reasonable approximation
for the description of natural phenomena. A function

$Q(x)$ which becomes zero exactly at $x = 0$ can hardly
have a real meaning, as this would correspond to an unstable initial mixture which would explode at the slightest impulse.

Under real conditions, $Q(x)$ either becomes zero
at $x > 0$ or, as in the most important instance of thermal
propagation, is never exactly zero but becomes negligibly
small already at $x > 0$. Under these conditions, the
uniqueness of the value of μ for which the solution
satisfies the boundary conditions is not subject to doubt.

Thermal Propagation of the Flame

The characteristic feature of the practically
most important case of thermal propagation of flames is
the unusually rapid increase of the rate of the reaction
with the temperature. On the sole basis of only this property of $Q(x)$ one can, according to Zeldovich[5], get an
approximate general form of solution, independent of the
concrete form of the function $Q(x)$, provided it possesses
this basic property.

To this end, we shall split the integration range
into two zones: a range adjacent to $x = 0$ where the re-
active velocity is negligible, and a range close to $x = 1$
where, on the contrary, it is very great. But, in the
latter case, in order to satisfy (VIII, 3), we must have
very large values of y, and then the first term, being
quadratic in y, will be markedly predominant over the
second term. In immediate vicinity of $x = 1$, y will
very rapidly fall to zero, but at the same time $\frac{dy}{dx}$ will
increase strongly.

Consequently, we can disregard the third term in
equation (VIII, 3) in the low-temperature zone and the
second in the high-temperature zone. The equation will then
take the form

$$y \frac{dy}{dx} = \mu y \qquad \text{(VIII, 7)}$$

close to $x = 0$, and

$$y \frac{dy}{dx} = - Q(x) \qquad \text{(VIII, 8)}$$

close to $x = 1$. The solution of (VIII, 7) has the form

$$y_1 = \mu x \qquad \text{(VIII, 9)}$$

and the solution of (VIII, 8) is of the form

$$y_2 = \sqrt{- 2 \int Q(x)\, dx} \qquad \text{(VIII, 10)}$$

In order to satisfy the boundary condition at $x = 1$, we can make the lower limit equal to 1. Transposing the integration limits, we get

$$y_2 = \sqrt{2 \int_x^1 Q(x)\, dx} \qquad \text{(VIII, 11)}$$

The equations (VIII, 9) and (VIII, 11) should satisfy the transition condition at the boundary between the zones $y_1 = y_2$.

Accurate location of this boundary is indefinite in principle. But, because of the sharp increase of the rate of the reaction with the temperature, its value at any temperature can be considered negligibly small as compared with its value at a higher temperature. Consequently, we can consider the expression (VIII, 9) valid up to temperatures immediately contiguous to the maximum combustion temperature, i.e., up to values of x close to unity. Hence, in (VIII, 9), we can put the transition point at

$$x \approx 1$$

On the other hand, in the integral (VIII, 11), one can place the lower limit as low as one wishes, as $Q(x)$ becomes negligibly small at low temperatures, and the value of the integral does not depend any more on the lower limit.

The transition condition then takes the form

$$\mu \approx \sqrt{2 \int_0^1 Q(x) \, dx} \qquad\qquad (VIII, 12)$$

which gives immediately the needed value of μ at which the solution of equation (VIII, 3) satisfies the boundary conditions (VIII, 6) for this limiting case of a very strong temperature dependence of the reaction velocity.

In the calculation of the integral in the right-hand member of (VIII, 12), the value of $Q(x)$ at the lower limit of integration should be considered negligibly small. The graphic approximate method is illustrated in

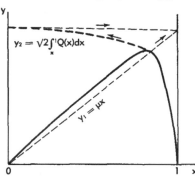

FIGURE 21.   APPROXIMATE METHOD OF SOLUTION OF THE EQUATION OF THERMAL PROPAGATION OF THE FLAME

Ordinates are

$$y = \frac{dx}{d\xi}$$

Abscissas are x. The full line represents the exact, and the dotted lines the approximate solution.

Figure 21. The dotted line represents the approximate, and the full line the exact solution.

If one assumes the simplest form of reaction kinetics, characterized by a definite order p, and uses the method of development of the exponent, as was done in formula (VI, 45), the function $Q(x)$ will take the form

$$Q(x) = n^p e^{-\dfrac{E}{RT_m^2}(T_m - T)} \tag{VIII, 13}$$

where n is the relative concentration of the reactant.
Expressing the latter, on the basis of the similitude of
the fields of concentrations and temperatures, by the
temperature according to (VI, 43), we get

$$Q(x) = \left(\dfrac{T_m - T}{T_m - T_o}\right)^p e^{-\dfrac{E}{RT_m^2}(T_m - T)} \tag{VIII, 14}$$

or, introducing the auxiliary variable x, according to
(VIII, 1)

$$Q(x) = (1 - x)^p e^{-\theta_m(1 - x)} \tag{VIII, 15}$$

where θ_m is the maximum dimensionless flame temperature

$$\theta_m = \dfrac{E}{RT_m^2}(T_m - T_o) \tag{VIII, 16}$$

Substituting (VIII, 15) in (VIII, 12) and considering that
the value of $Q(x)$ at the limit of integration

$$Q(0) = e^{-\theta_m} \ll 1$$

should be negligibly small, we get after integration

$$\mu = \sqrt{2\,\dfrac{p!}{\theta_m^p + 1}} \tag{VIII, 17}$$

With the aid of (VIII, 4) we get for the flame propagation
velocity

$$w = \sqrt{\frac{2p!}{\theta_m^p + 1} \frac{a}{\tau_m}} \qquad \text{(VIII, 18)}$$

This result has already been given in chapter VI, formulas (VI, 47) and (VI, 49).

Development and application of this theory were given in the articles of Zeldovich and of Semenov[6,12].

Method of the Heat Flux

Instead of resorting to an integration of the differential equation, one can obtain the main results of the theory of thermal propagation of flames in a very simple and direct way by considering the heat flow from the flame zone. Let us designate by y the temperature gradient at the flame front. The heat flow from the flame zone will then be expressed by

$$q = - \lambda y_m \qquad \text{(VIII, 19)}$$

where λ is the heat conductivity of the combustible mixture, and y_m the maximum value of the temperature gradient at the flame front. This heat flow is spent on heating the combustible mixture from the initial temperature T_0 to the combustion temperature T_m,

$$q = c\rho w(T_m - T_0) \qquad \text{(VIII, 20)}$$

where c is the heat capacity of the combustible mixture, ρ its density, w the flame propagation velocity. To calculate y_m accurately, it is necessary to use, in the flame zone, equation (VIII, 8); this will give a result identical with (VIII, 12). Or else, one can make use of the concept of the flame front thickness. We introduce the auxiliary magnitude ξ, which we shall term

the thermal thickness of the flame front, and define it
so that

$$y_m = - \frac{T_m - T_o}{\xi_1}$$ (VIII, 21)

The physical meaning of that magnitude is that the true
temperature distribution along the flame front which has
the form of a continuous curve, is replaced by a broken
line, obtained by drawing a tangent to the true curve at
the inflection point. Combining the expressions (VIII, 19)
and (VIII, 21), we get for the flame propagation velocity

$$w = \frac{\lambda}{c\rho\xi_1} = \frac{a}{\xi_1}$$ (VIII, 22)

where a is the temperature conductivity of the combustible
mixture. On the other hand, the thickness of the flame
propagation front is linked with the velocity of propagation. We shall term chemical thickness of the flame front
the thickness of the zone in which the chemical combustion
reactions in the flame front would take place at the maximum temperature of combustion in the absence of diffusion
and heat transfer effects. The chemical thickness of the
flame front is related with the previously introduced reaction time by

$$\xi_2 = w\tau_m$$ (VIII, 23)

The thermal flame front thickness ξ_1 is the thickness of
the zone in which there is a substantial change of temperature, whereas the chemical thickness ξ_2 is the thickness
of the zone in which the chemical process takes place. It
is evident that the thermal flame front thickness cannot
under any circumstances be less than the chemical thickness, since evolution of heat and, consequently, rise of
temperature, is inevitable wherever chemical combustion

reactions occur. Actually, the chemical flame front thickness is always smaller than the thermal, as the combustion
zone is broadened as a result of diffusion and heat transfer.

We shall put

$$\xi_2 = F\xi_1 \qquad\qquad \text{(VIII, 24)}$$

where F is a dimensionless factor smaller than unity,
the numerical value of which is determined by the concrete
form of the kinetics of the chemical combustion reactions,
i.e., by the dependence of the rate of reaction on the
temperature and the reactant concentrations. Combining
(VIII, 22) - (VIII, 24), we get finally for the flame
propagation velocity

$$w = \sqrt{F\,\frac{a}{\tau_m}} \qquad\qquad \text{(VIII, 25)}$$

This formula coincides with (VI, 47) or (VIII, 4) if in
the latter one puts $\mu = \sqrt{F}$. We have now clarified the
physical meaning of that magnitude. The dimensionless
flame propagation velocity is nothing but the square root
of the ratio of the chemical and the thermal flame front
thickness.

Diffusional (Chain) Flame Propagation in Second-order Auto-catalysis

By diffusional, or chain propagation of the flame
we shall mean a process wherein the propagation of the
flame is linked not with transfer of heat but with diffusion
of the active product of the autocatalytic (chain) reaction. This concept must not be confused with the terms
of "diffusional combustion" sometimes encountered in the
literature, and which refers to the case when the components of the combustible mixture are not mixed in advance
and the velocity of the combustion is determined by their

mutual diffusion. In diffusional, or chain combustion,
the form of the functions $Q(x)$ in (VIII, 3) is determined by the kinetics of the reaction; nothing general
can be said about it. In principle, various forms of the
function $Q(x)$ are possible, but an analytical solution
of equation (VIII, 3) and a calculation of the value of
μ at which this solution satisfies the boundary conditions
is practically not possible.

We[5] have examined the simple case when equation
(VIII, 3) does have a simple analytical solution. This is
the case of second-order autocatalysis. In order to obtain a simple analytical solution of equation (VIII, 3),
we shall reverse the problem. Instead of seeking, from
the prescribed function $Q(x)$, the corresponding form of
the function $y(x)$ which satisfies the boundary conditions,
we shall seek that form of $Q(x)$ for which the chosen
$y(x)$ will satisfy the equation.

As the simplest form of the function $y(x)$
satisfying the boundary conditions (VIII, 6) we shall take

$$y = x(1 - x) \qquad\qquad \text{(VIII, 26)}$$

Substituting this expression for y in equation (VIII, 3)
we get

$$- 2x^2(1 - x) + (1 - \mu)x(1 - x) + Q(x) = 0 \qquad \text{(VIII, 27)}$$

In this way, the simplest form of solution of equation
(VIII, 3) will be found if the function $Q(x)$ is given
in the form

$$Q(x) = 2[x^2(1 - x) + ax(1 - x)] \qquad \text{(VIII, 28)}$$

The function $Q(x)$ has at $x = 0$ a negative derivative,
and hence the solution ought to be unique.

The corresponding expression for the rate of

reaction will be

$$\frac{dx}{dt} = \varphi x^2 (1 - x) - a\varphi x(1 - x) \qquad \text{(VIII, 29)}$$

where

$$\varphi = \frac{2}{\tau_m} \qquad \text{(VIII, 30)}$$

This form of kinetics corresponds to autocatalysis of the second order with a rate constant φ, concomitant on a simultaneous first-order reaction of consumption of the active product. The rates of both the first and the second reaction are proportional to the concentration of the original substance. The constant a is the ratio of the rate constants of the second and first reaction. In the terminology of Semenov which is accepted in the chain theory, we should term the first reaction the quadratic branching or chain interaction, and the second, the chain termination process. The constant a is the ratio of the rate constants of the chain termination and the chain branching reactions. Combining the expressions (VIII, 27) and (VIII, 28) we conclude that the dimensionless flame propagation velocity μ is related with the constant a characteristic of the kinetics of the reaction by

$$\mu = 1 - 2a \qquad \text{(VIII, 31)}$$

With the aid of (VIII, 5) and (VIII, 30) we get for the flame propagation velocity

$$w = (1 - 2a) \sqrt{\frac{\varphi D}{2}} \qquad \text{(VIII, 32)}$$

This formula was used by Voronkov and Semenov[7] in the treatment of experimental data of propagation of the cool flame in very lean mixtures of carbon disulfide with air.

Combustion in a Moving Gas

We have considered the so-called fundamental velocity of propagation of the flame, i.e., the velocity of propagation relative to a motionless gas. Actually, the combustion process is always accompanied by a motion of the gas. Even if the motion is not brought about artificially, it inevitably will arise spontaneously as a result of thermal expansion.

The curvature of the flame front, linked with the motion of the gas, leads to an increase of the velocity of the combustion. Let us consider a plane frame front of surface area σ moving in a direction perpendicular to its plane with the velocity v. The volume of gas burning per unit time will be expressed by

$$w = v\sigma \qquad\qquad (VIII, 33)$$

Let the surface area of the curved flame front be S. At each point of the front, the flame spreads perpendicularly to its surface with a velocity equal to the fundamental velocity w. Consequently, the same volume of gas mixture burning per unit time will be expressed as

$$w = wS \qquad\qquad (VIII, 34)$$

Combining the expressions (VIII, 33) and (VIII, 34) we get

$$v = w \frac{S}{\sigma} \qquad\qquad (VIII, 35)$$

Formula (VIII, 35) expresses the so-called law of areas. According to this law, when the flame front is curved, the velocity of propagation increases proportionally to the increase of its surface area.

It is easily remarked that the cosine of the angle between the perpendicular to the surface of the flame

and the direction of the gas flow is equal to the ratio of
the fundamental velocity w and real velocity of combustion v (cosine law).

Turbulent Combustion

Combustion in a turbulent gas flow has a very
great technical importance. Turbulence increases the
flame propagation velocity and permits very strong intensification of the combustion process.

The mechanism of the acceleration of the combustion by turbulence can be conceived in two ways. On
the one hand, one can assume that turbulence increases
the velocity of heat transfer in the flame front, without influencing the course of the chemical combustion reaction proper. This view is the only possible one if one
assumes, as is usually done in the literature[8,9], that
the combustion process takes place in a perfectly homogeneous medium, ideally mixed in advance. Practically it
is more commonly found that the combustion process takes
place simultaneously with the mixing of the fuel and the
air. If the fuel and the air are separated by a continuous boundary, we shall observe the usual diffusion combustion obeying the usual laws of diffusion, or mixing of
the streams. The theory of such a process was discussed
in the literature[10].

In our work with Minskii[11] we first discussed
the particular instance of the turbulent combustion process
when the fuel is fragmented into discrete small volumes
distributed in the air stream. The velocity of the combustion is determined by the law of micromixing, i.e.,
mutual mixing of the small volumes of the fuel and the
surrounding air. This process we termed "micro-diffusion-
al turbulent combustion". In it, the acceleration of the
combustion by turbulence is due, in the first place, to an
acceleration of the chemical reactions at the flame front.
The rate of these reactions is determined, in this instance,

by the micromixing process which is accelerated by turbulence. Microdiffusional turbulent combustion is usually
localized in the turbulent wake behind some obstacles to
the flow of the combustible mixture. Such bodies serve as
stabilizers of the flame; the flame clings to them and
can remain stable only in close contact with them.

In our opinion whenever turbulence is found to
accelerate combustion markedly, the process is actually
microdiffusional combustion, with intensification of the
heat transfer at the flame front playing only a subordinate
role. Further development of the concept of microdiffusional combustion has led us to extend to this field, the ideas
of the diffusional and kinetic ranges of reaction which
have proved so fruitful in diffusional kinetics of heterogeneous reactions.

Inasmuch as advance mixing of the fuel and the
air can never be ideal, combustion can practially always
follow a microdiffusional course, if the velocity of the
micromixing is small as compared with the rate of the
chemical reactions proper. We can, by appropriately con-
trolling the conditions, improve a course in which either
diffusional or kinetic factors play the predominant role
in analogy to the alternative possibility for heterogeneous combustion to take place either in the diffusional
or in the kinetic region. Only under definitely prescribed
conditions is it possible to state what degree of perfec-
tion of the preliminary mixing is necessary for the mixture to be considered "homogeneous" with respect to the
combustion process.

The acceleration of the combustion by turbulence
cannot be unlimited; a limit to the acceleration is set
by the transition of the combustion reaction into the
kinetic region.

The concept of microdiffusional combustion provides the most natural interpretation of the proportionality

between the velocity of the combustion and the pulsation velocity or the flow velocity, which is the most characteristic feature of turbulent combustion.

Literature

1. NUSSELT, Z. Ver. Deutsch. Ing. 59, 872 (1915).

2. JOUGUET, Compt. rendus 156, 872 (1913); 168, 820 (1919).

3. DANIELL, Proc. Roy. Soc. A 156, 393 (1930).

4. KOLMOGOROV, PETROVSKII, PISKUNOV, Byull. Moskov. Gos. Univ. Sect. A 1, No. 6 (1937).

5. ZELDOVICH and FRANK-KAMENETSKII, Doklady Akad. Nauk S.S.S.R. 19, 693 (1938).

6. ZELDOVICH, Teoriya goreniya i detonatsii gazov (Theory of combustion and detonation of gases), Moscow 1944.

7. VORONKOV and SEMENOV, Zhur. Fiz. Khim. 13, 1695 (1939).

8. DAMKOHLER, Z. Elektrochem. 46, 601 (1940).

9. SHCHELKIN, Zhur. Tekh. Fiz. 13, 520 (1943).

10. BURKE and SCHUMANN, Ind. Eng. Chem. 20, 998 (1928).

11. MINSKII and FRANK-KAMENETSKII, Doklady Akad. Nauk S.S.S.R. 50, 353 (1945).

12. ZELDOVICH and SEMENOV, Zhur. Eksptl. Teoret. Fiz. 10, 1116 (1940).

CHAPTER IX: THERMAL REGIME OF HETEROGENEOUS EXOTHERMAL REACTIONS

We have considered cases where either diffusion (diffusional kinetics) or evolution and propagation of heat (thermal kinetics) is essential for the course of a chemical reaction.

We shall now turn to processes in which both heat transfer and diffusion are essential at the same time. One meets with such processes when one analyzes the thermal regime of heterogeneous reactions. If the heterogeneous reaction has a significant thermal effect, the temperature of the surface at which the reaction takes place is different from the temperature in the volume of the gas flow and from the temperature of the surrounding medium.

The quantity of heat evolved at the surface is determined by the macroscopic velocity of the reaction and, specifically in the diffusional region, by the velocity of diffusion. The quantity of heat removed from the surface is determined by the conditions of heat exchange. Thus, the stationary temperature of the surface which is established when the rate of removal of heat balances the rate of supply, depends on the ratio of the rate of reaction and the intensity of heat exchange. To calculate that stationary temperature, it is necessary to take into account both diffusion and heat transfer processes. In the solution of the problem, we shall evidently need

the equations of heat conductance and of diffusion for
simultaneous occurrence of these processes, discussed in
chapter IV. In particular, thermal diffusion phenomena
may prove essential.

For reasons already discussed in the foregoing
chapters, most important from the point of view of the
thermal regime are exothermal processes. An endothermal
reaction is self-regulating from the thermal point of
view and always leads to a stable thermal regime. In contrast, in an exothermal reaction, the thermal regime can
become unstable; there is the possibility of critical
phenomena of breakdown of the regime, of transition from
one thermal regime to another, as was first discussed in
our work[1].*

The ideas and methods which have been used in
the theory of thermal ignition in homogeneous reactions
will now be applied to the problem of the thermal regime
of heterogeneous exothermal reactions. The difference is
that in the heterogeneous case the rate of reaction can
no longer increase without limit, up to highest temperatures. The rate of a heterogeneous chemical process is
determined both by the true rate of the chemical reaction
at the surface and by the rate of supply of reactants to
that surface through molecular or convective diffusion.
At low temperatures, as long as the rate of reaction is
small as compared with the velocity of diffusion (kinetic
region), the overall rate of the process is determined by
the true kinetics at the surface and increases exponentially with the temperature according to Arrhenius' law.
But this increase can go on only until the velocity of the
reaction becomes comparable with the velocity of the
diffusion. From then on the process will pass into the
diffusional region where its rate is always determined by
the velocity of diffusion, and increases only very slowly

---

* See also: Wagner, C., Die chemische Technik 18 1, 28
(1945).

with the temperature. With such a temperature dependence
of the velocity of evolution of heat and under definite
conditions of dissipation of heat, three stationary thermal states are possible; the intermediate state is unstable, the upper corresponds to reaction taking its
course in the diffusional, and the lower to reaction in
the kinetic region. Ignition of the surface represents
a discontinuous transition from the lower to the upper
stationary thermal regime. The reverse transition from
the upper to the lower regime also takes place discontinuously at the critical condition of extinction which
does not coincide with the condition of ignition.

While the condition of ignition in heterogeneous
reactions is entirely analogous to that for homogeneous
reactions, extinction represents an essentially new
phenomenon observable only in stationary processes. Analogous phenomena in homogeneous reactions are possible only
with entirely special methods of realization, and were
predicted by Zeldovich[15] after our theory had been
developed.

Determination of critical conditions of ignition
and extinction can serve as a method of study of the kinetics of strongly exothermal heterogeneous reactions, the
rate of which can not easily be measured directly.

The temperature rise at the surface in the upper
thermal regime is entirely determined by diffusion phenomena. If the diffusion coefficient of the reacting gas
is not different from the temperature conductivity of the
gaseous mixture, the temperature of the surface will be
equal to the theoretical temperature of combustion. This,
however, is in no way a consequence of the law of conservation of energy, as in this case we have not a closed system but a stationary process. Differences between diffusion
and temperature conductivity coefficients, and also thermal
diffusion phenomena, can cause a marked difference between
the temperature of the surface and the theoretical

temperature. Specifically, if the molecular weight of the
reacting gas is less than the mean molecular weight of the
gaseous mixture, the temperature of the surface is higher
than the theoretical temperature of combustion, which was
confirmed experimentally in the thesis of Buben[3] for the
catalytic oxidation of hydrogen on platinum.

We have[1] applied our theory of the thermal
regime of exothermal heterogeneous reactions to the most
important instance of heterogeneous combustion, namely to
the combustion of coal. The processes of oxidation and
combustion of coal observed by Grozdovskii and Chukhanov[12]
and treated by these authors as two different chemical reactions, could be interpreted as two distinct thermal
regimes of one and the same reaction. Determinations of
the conditions of ignition of carbon wires in a flow of
air have permitted a number of conclusions on the kinetics
of the reaction between carbon and oxygen at high temperatures[18].

Application of the theory to the technically
important process of catalytic oxidation of isopropyl
alcohol to acetone has shown ways of perfecting this process[9]. Our theory also was applied, by Bresler and
Zinovev[10], to the construction of electric gas analyzers,
based on the principle of measurement of the temperature
rise of a catalyzing surface.

Qualitative Theory of the Phenomena of Ignition and Extinction Regardless of the Kinetics of the Reaction

We shall examine the thermal regime of the surface in the general case where the kinetics at the surface
is immaterial; it is necessary only that the rate of reaction increase with the temperature monotonously and
sufficiently rapidly. We shall use the method applied by
Semenov to the theory of thermal ignition in homogeneous
reactions (see chapters VI and VII).

In Figure 22, we plot on the abscissa, the
temperature, and on the ordinate the rates of supply and
removal of heat. Curve 1
in the figure represents
the rate of the reaction or
the rate of supply of heat
proportional to the rate of
reaction, i.e., the amount
of heat evolved on unit surface area per unit time.
The portion A of the curve
corresponds to the kinetic
range; here the rate of reaction increases exponentially with the temperature
and is independent of the
velocity of flow. The portion B of the curve corre-

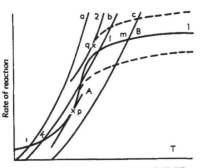

FIGURE 22. STATIONARY THERMAL REGIME OF THE
SURFACE

Ordinates are rates of reactions. Abscissas are
temperatures. The thick line represents the
temperature dependence of the rate of the heterogeneous reaction. The dotted lines in the diffusional region correspond to different rates of
flow of the gas. The thin lines represent the
rate of removal of heat. The meaning of the
points of contact and of intersection is explained
in the text.

sponds to the diffusional range where the rate of reaction
increases with the temperature only very weakly but depends strongly on the velocity of flow. The different
dotted lines in portion B refer to different velocities of
the gas flow. The greater is that velocity, the higher
lies the corresponding curve. The curves 2 in Figure 22
represent the intensity of removal of heat from the surface
under different conditions, expressed in the same units.
Inasmuch as the loss of heat by radiation, too, depends on
the temperature much less than the rate of reaction, the
curves 2 over short temperature intervals can be considered rectilinear.

Exact allowance for the effect of radiation on
the form of these curves does not affect the qualitative
results. The abscissas in Figure 22 represent the temperature at which the reaction takes place, i.e., the temperature of the gas at the surface, which we shall consider the
same as the temperature of the surface itself (accommodation
effects, essential only at very low pressures, will not be

taken into consideration here). The position of curve 2
(the point of its intersection with the axis of abscissas)
is determined by the temperature of the medium in which
heat is exchanged. This temperature we shall consider to
be the same as the temperature of the gas in the flow, at
a distance from the surface, where diffusion of the reactant takes place. The case of direct loss of heat by
radiation into a medium colder than the reacting gas can
be easily discussed by the same method, but this will not
be done here.

Depending on the mutual position of curves 1 and
2, i.e., on the ratio of the rate of reaction and the intensity of heat exchange, the thermal regime of the gas
at the surface and, consequently, of the surface itself,
will be different. The stationary temperature rise at
the surface is evidently determined by the point of intersection of the curves 1 and 2. Examination of these points
of intersection for different positions of the curves
leads immediately to a series of very interesting conclusions.

If curve 2 is in position a, only stationary
state (point i) is possible, corresponding to small temperature rises in the kinetic region. In position b, three
stationary states are possible of which, however, only
the upper and the lower are stable, the intermediate
state being unstable. Which stationary state will become established in this region, depends evidently on the
initial state of the surface. If, originally, the surface
was at a high temperature, and is placed in a cooler gas,
its temperature will drop only as far as the temperature
corresponding to the upper intersection (point l) and will
not fall any further, i.e., the upper stationary state
will be established.

If curve 2 is in position c, only one stationary
state (point m) is possible. This state corresponds always
to the diffusional region and, at sufficiently great

velocity of the gas flow or small rate of heat exchange,
to a large local temperature rise. The boundaries between
these three regions will correspond to conditions where
curve 2 is in contact with curve 1. Such a contact can
take place at two points, designated in Figure 22 by p
and q. If an originally cool surface is placed in a reacting gaseous medium, and the parameters which affect the
rate of reaction on the conditions of the heat exchanged,
the temperature rise at the surface and the macroscopic
rate of the process will change continuously as long as
we remain within the bounds of the regions a and b. On
reaching the point of contact p of curves 1 and 2, a more
or less sharp change of the temperature rise and of the
observed rate of reaction will occur: we shall pass discontinuously from the type k to the type in intersection.
Consequently, the condition of contact of the curves of
supply and loss of heat at the point p is the critical
condition of ignition of the solid body. If, in particular, the temperature in the space of the gas flow is
changed with other conditions kept unchanged, curve 2
will shift with curve 1 remaining in the same position.
The temperature at which it will come in contact with
curve 1 at the point p, is the temperature of ignition of
the solid surface.

        If an originally incandescent surface is placed
in a cooler medium, the passage through point p will be
entirely uncritical. The transition between regions c
and b will be entirely continuous and uninterrupted, and
the upper stationary state, corresponding to the diffusional region (point m or l) will prevail throughout. But if
the intensity of removal of heat is increased, with curve
2 coming into contact with 1 at the point q, there will be
an abrupt change of the temperature of the surface, and we
shall pass to the lower stationary state (point i) corresponding to a very small local temperature rise in the
kinetic region.

The condition for the contact of curves 2 and 1
at the point q is the critical condition of extinction
of a hot surface. We see that it is not identical with
the condition of ignition.

In this way, ignition of a solid body is identically tied with the passage of the reaction from the kinetic into the diffusional region. Conversely, in the case
of a sufficiently strongly exothermal reaction, the diffusional is inevitably also a region of strong temperature
rise, increasing with the velocity of the gas flow. This
result is perfectly natural. The diffusional range is
the region in which the concentration of the reactant
differs from the concentration at the surface. It is
natural that in this region the temperature at the surface should also be different from the temperature in the
volume. The transition from the kinetic into the diffusional region and vice versa takes place discontinuously at the
critical conditions of ignition and extinction; the jump
is the greater the faster is the gas flow. The intermediate range between the diffusional and the kinetic regions is absent in this case, as it corresponds to an unstable thermal regime.

Summing up the above, we come to the conclusion
that in an exothermal heterogeneous reaction there can be
two stationary thermal states, one corresponding to a small
temperature rise in the kinetic region, the other to a
large temperature rise in the diffusional region.

We shall refer to the first as the lower, and to
the second as the upper thermal regime. Passage from one
regime to the other occurs discontinuously at the critical
conditions of ignition and extinction.

Mathematical Theory of the Phenomena of Ignition and Extinction for Reactions of the First Order

We shall consider the simplest case when the

true kinetics at the surface follows the first order[2].
The macroscopic reaction rate is then expressed by formula (II, 5)

$$q = \frac{k\beta}{k + \beta} \; C$$

where q is the amount of substance reacting on unit surface area per unit time, C is the concentration of the reactant in the volume, k is the rate constant of the chemical reaction at the surface, depending on the temperature according to Arrhenius' law

$$k = ze^{-\frac{E}{RT}} \qquad\qquad (IX, 1)$$

and β is the diffusion velocity constant, defined by the relation

$$\beta = \frac{Nu \; D}{d} \qquad\qquad (IX, 2)$$

where Nu is the Nusselt criterion depending on the geometric configuration of the system and the conditions of convection, D is the diffusion coefficient of the reactant, and d the characteristic linear dimension of the system.

With Q designating the heat effect of the reaction, the amount of heat evolved on unit reaction surface per unit time is

$$q_1 = Qq = Q \frac{k\beta}{k + \beta} \; C \qquad\qquad (IX, 3)$$

A stationary thermal state will be established when this supply of heat becomes equal to the amount of heat removed from unit surface area per unit time. The latter amount can be written

$$q_2 = \alpha(T - T_0) \qquad\qquad (IX, 4)$$

where α is the heat exchange coefficient, and T_0 the temperature of the surrounding medium into which the heat is transferred.

Thus, the stationary temperature of the surface can be found by solving the equation

$$Q \frac{ze^{-\frac{E}{RT}} \beta}{ze^{-\frac{E}{RT}} + \beta} C = \alpha(T - T_0) \qquad\qquad (IX, 5)$$

We shall seek an approximate solution on the assumption $E \gg RT$. In this case, one can make use of the development of the exponent in Arrhenius' law which we have applied in the theory of thermal ignition for homogeneous reactions (see chapter VI)

$$k = ze^{-\frac{E}{RT}} \approx ze^{-\frac{E}{RT_0}} \cdot e^{\theta} \qquad\qquad (IX, 6)$$

where

$$\theta = \frac{E}{RT_0^2} (T - T_0)$$

is the dimensionless temperature.

The results obtained will be accurate only at $T - T_0 \ll T_0$, i.e., at small temperature rises. They therefore will be sufficiently accurate for the condition of ignition; for the condition of extinction, which can occur at sizeable temperature rises, the accuracy of our calculation will be lower.

Using the development (IX, 6) and transforming

equation (IX, 5) to a dimensionless temperature, we put it in the form

$$\frac{Q}{\alpha} ze^{-\frac{E}{RT_o}} \frac{E}{RT_o^2} C \frac{e^\theta}{e^\theta \frac{ze^{-\frac{E}{RT_o}}}{\beta} + 1} = \theta \qquad (IX, 7)$$

The transcendent equation (IX, 7) defines the value of θ corresponding to a stationary temperature rise of the surface. For the dimensionless parameters in equation (IX, 7) we introduce the notation

$$\frac{Q}{\alpha} \frac{E}{RT_o^2} ze^{-\frac{E}{RT_o}} C = \delta$$

$$\frac{ze^{-\frac{E}{RT_o}}}{\beta} = \mu$$

As the equation contains only these two dimensionless parameters, the critical conditions of ignition and of extinction must be of the form

$$\delta_{cr} = F(\mu)$$

The first of these parameters is entirely analogous to the parameter introduced in chapter VI under the same designation in the theory of thermal ignition in homogeneous reactions.

In the notation just adopted, equation (IX, 7) will be written

$$\delta \frac{e^\theta}{\mu e^\theta + 1} = \theta$$

or

$$\delta = \theta(\mu + e^{-\theta}) \qquad\qquad (IX, 8)$$

In order to elucidate the general properties of the solutions of equation (IX, 8) we designate its right-hand member by $f(\theta)$. Clearly, if $f(\theta)$ is a monotonous function, the equation will always have a unique solution, and critical conditions will not occur. If, however, $f(\theta)$ has an extreme value, there can be several stationary states, the extreme points corresponding to the critical conditions of transition from one state to another.

Thus, the critical conditions of ignition and extinction should be found from the equation

$$\frac{df(\theta)}{d\theta} = \mu + e^{-\theta} - \theta e^{-\theta} = 0$$

$$(IX, 9)$$

$$\mu = e^{-\theta}(\theta - 1)$$

Solution of this equation will give the extreme values of θ as a function of μ. Substituting such a value of θ in equation (IX, 8) we get the critical value of δ as a function of μ.

Consequently, the critical conditions of ignition and extinction

$$\delta_{cr} = F(\mu)$$

are given in parameter form by equations (IX, 8) and (IX, 9). Substituting the value of μ from (IX, 9), we get the final form of the critical condition in parameter form

$$\delta_{cr} = \theta^2 e^{-\theta}$$

$$(IX, 10)$$

$$\mu = e^{-\theta}(\theta - 1)$$

Inasmuch as μ is a positive magnitude, θ can take all values from 1 to ∞. Prescribing different values of θ within these limits and substituting them in (IX, 10), we get the relation between δ and μ corresponding to the critical condition.

The two parameters δ and μ which we have introduced depend on the rate of the reaction and, consequently, depend on the temperature exponentially. Therefore, for concrete calculations, it is more convenient to introduce a new parameter

$$\xi = \frac{\delta}{\mu} = \frac{Q}{\alpha} \frac{E}{RT_o^2} \beta C \qquad (IX, 11)$$

which does not contain the rate of reaction, and to express the critical condition as a dependence of the critical value of δ on the parameter ξ. In order to make the numerical calculations still more convenient, we shall further introduce, instead of θ, the auxiliary variable $x = e^{-\theta}$, varying from zero to $\frac{1}{e}$. The critical condition in parameter form will take the form

$$\delta_{cr} = x\ln^2 x$$

$$\xi = \frac{\ln^2 x}{-\ln x - 1} \qquad (IX, 10a)$$

Prescribing different values of x from 0 to $\frac{1}{e}$, we get the range of critical temperature phenomena represented graphically in Figure 23. The general properties of this range are easily found by examination of equation (IX, 9) which defines the extreme values of θ. The right-hand member of this equation as a function of θ has a maximum at $\theta = 2$, with the corresponding value $\mu = \frac{1}{e^2}$. Consequently, at $\mu < \frac{1}{e^2}$, equation (IX, 9) has two solutions, and at $\mu = \frac{1}{e^2}$ none. Correspondingly, at

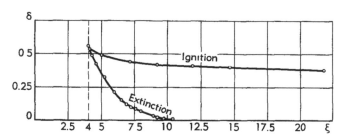

FIGURE 23.   CRITICAL CONDITIONS OF IGNITION AND EXTINCTION
FOR A SECOND-ORDER REACTION

Ordinates are the critical values of the parameter δ.
Abscissas give the values of the parameter ξ. The upper
curve represents the critical condition of ignition, and
the lower the critical condition of extinction.

$\mu < \dfrac{1}{e^2}$, the function $f(\theta)$ has two extreme values, one
maximum and one minimum; at $\mu > \dfrac{1}{e^2}$ it has none.

Figure 24 represents the form of the function
$f(\theta)$ at $\mu < \dfrac{1}{e^2}$, when $f(\theta)$ has a maximum and a minimum.

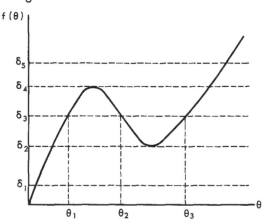

FIGURE 24.   GRAPHIC SOLUTION OF EQUATION (IX, 8)

Ordinates are $f(\theta)$. Abscissas are θ. The
horizontal dotted lines represent the different
values of the parameter δ.

With a given value of δ, the solutions of equation (IX, 8)
correspond to the intersection of the curve $f(\theta)$ with
the straight line running parallel to the axis of abscissas
at the distance δ. At small values of δ (δ_1 on

Figure 24) there is only one solution, corresponding to a small temperature rise (lower temperature regime). At a greater $\delta = \delta_5$ there is also only one solution, corresponding to a large temperature rise (upper temperature regime). At intermediate values $\delta = \delta_3$, three solutions are possible, of which θ, corresponds to the lower temperature, θ_3 to the upper temperature, and θ_2 to the intermediate unstable temperature regime.

If, starting from a low temperature, we increase δ gradually, the low temperature regime will maintain itself until the value $\delta = \delta_4$ is reached; at higher δ the low temperature regime is impossible, and there must be a discontinuous jump to the upper temperature regime. Consequently, the condition $\delta = \delta_4$ gives the critical condition of ignition.

If, starting at a high temperature, one decreases δ, the upper temperature regime will maintain itself as far as the value $\delta = \delta_2$ which gives the critical condition of extinction. As we see, the critical conditions actually correspond to the extreme values of the function $f(\theta)$.

If $f(\theta)$ has no extreme value, equation (IX, 8) will always have one solution, then there will be no critical phenomena.

Thus, in order for critical phenomena to occur, it is necessary that $\mu < \dfrac{1}{e^2}$. By equation (IX, 10) the corresponding value of δ is $\dfrac{4}{e^2}$; consequently, critical conditions are possible only at $\xi > 4$, as can also be seen from Figure 23. At $\xi < 4$, the temperature of the surface changes continuously with the external conditions. On the contrary, at $\xi > 4$, one should observe a discontinuous jump from the diffusional to the kinetic region, and vice versa, at the critical conditions of ignition and extinction.

If the thermal effect and the activation energy
are high, the parameter ξ will also be very large. We
shall examine more closely this most interesting limiting
case where the critical phenomena are most pronounced.

From (IX, 10a) it can be seen that at $\xi \gg 1$
the critical condition can correspond either to
$\theta = (- \ln x) \approx 1$ or to $\theta = (- \ln x) \gg 1$. In the first
case we get

$$\delta_{cr} = \frac{1}{\theta} \qquad \qquad (IX, 12)$$

This is the limiting form of the condition of ignition at
high values of ξ.

If, in (IX, 10a), one puts $(- \ln x) \gg 1$, one
will get the limiting form of the condition of extinction
at high ξ. Equation (IX, 10a) gives

$$\xi \approx - \ln x \qquad \qquad (IX, 13)$$

$$\delta_{cr} \approx \xi^2 e^{-\xi} \qquad \qquad (IX, 14)$$

For the temperature of the surface one gets

$$\theta = - \ln x \approx \xi \qquad \qquad (IX, 15)$$

This result is entirely natural. In fact, in a reaction
taking its course in the pure diffusional range, the rate
of evolution of heat is expressed by $Q\beta C$; equating this
rate to the rate of removal of heat $\alpha(T - T_0)$, we find
that in the diffusional range the temperature rise of the
surface is equal to

$$(T - T_0)_m = \frac{Q\beta C}{\alpha}$$

By the above definition of the dimensionless temperature

θ, this temperature rise in the dimensionless form will be

$$\theta_m = \frac{E}{RT_o^2} \frac{Q\beta C}{\alpha}$$

which coincides with the definition of the parameter ξ.

Thus, the parameter ξ represents the dimensionless maximum temperature of the surface, attainable in the purely diffusional region. The result (IX, 15) means that at high values of ξ the upper temperature regime corresponds to the pure diffusional region. It is easily shown that the lower temperature regime corresponds to the pure kinetic region.

$$(\mu = \frac{\delta}{\xi} \ll 1, \quad \text{i.e.,} \quad \beta \gg ze^{-\frac{E}{RT_o}})$$

The intermediate region at $\xi \gg 1$, i.e., in the case of strongly exothermal reactions with a large temperature coefficient, cannot be realized, as it corresponds to unstable thermal regimes.

We can transform the condition of extinction (IX, 14) so that it will include the temperature of the surface. To this end, we shall, with the aid of (IX, 15), write (IX, 14) in the form

$$\delta_{cr}e^\theta = \xi^2 \qquad \qquad (IX, 16)$$

We now can, to a large extent, eliminate the inaccuracy due to the use of the development (IX, 6). There we have replace

$$ze^{-\frac{E}{RT}} \quad \text{by} \quad ze^{-\frac{E}{RT_o}} \cdot e^\theta$$

We now can go back again to the accurate Arrhenius law. To this end, we substitute the value of δ in equation (IX, 16), and get the critical condition of extinction in the form

$$\frac{Q}{\alpha} \frac{E}{RT_o^2} \, Cze^{-\frac{E}{RT_o}} \cdot e^\theta = \xi^2$$

We now replace

$$e^{-\frac{E}{RT_o}} \cdot e^\theta \quad \text{by} \quad e^{-\frac{E}{RT}}$$

and get

$$\frac{Q}{\alpha} \frac{E}{RT_o^2} \, Cze^{-\frac{E}{RT}} = \xi^2$$

Substituting the value of ξ, we get after simplification

$$\frac{ze^{-\frac{E}{RT}}}{\beta} = \xi \qquad\qquad (IX, 17)$$

or

$$ze^{-\frac{E}{RT}} = \frac{E}{RT_o^2} \frac{Q\beta^2 C}{\alpha} \qquad\qquad (IX, 18)$$

where T is the temperature of the surface at which the extinction takes place, and T_o is the temperature of the surrounding medium.

The condition of extinction in the form (IX, 17) or (IX, 18) can be used without serious error also when the temperature rise of the surface at the limit of extinction is quite large, inasmuch as the error due to the use of the development (IX, 6) has been minimized thanks to the subsequent return to the accurate Arrhenius law.

If in the same way, one expresses the limiting

form of the critical condition of ignition at high ξ by the temperature of the surface, formula (IX, 12) will give

$$ze^{-\frac{E}{RT}} = \frac{RT_0^2}{E}\frac{\alpha}{QC}$$

At high values of ξ, at the ignition limit

$$\theta = 1; \quad e^{-\frac{E}{RT}} \approx e^{-\frac{E}{RT_0}} \cdot e$$

According to the above-given definition of the parameter ξ we have

$$\frac{RT_0^2}{E}\frac{\alpha}{QC} = \frac{\beta}{\xi}$$

$$\frac{E}{RT_0^2}\frac{Q\beta^2 C}{\alpha} = \beta\xi$$

Consequently, at high ξ we can represent the critical conditions of ignition and extinction in the following instructive and mnemonically easy form:

Condition of ignition

$$ze^{-\frac{E}{RT}} = \frac{\beta}{\xi}$$

Condition of extinction

$$ze^{-\frac{E}{RT}} = \beta\xi$$

where T is the temperature of the surface. Thus, at the ignition limit, the reaction rate constant is ξ times as small as the diffusion velocity constant, and at the extinction limit the reaction rate constant is ξ times as great as the diffusion velocity constant.

This theory refers to the simplest case of a first-order reaction and, besides, it is only approximate, inasmuch as it involves a development of the exponential.

The problem of the conditions of ignition and of extinction for any kinetics of the reaction was examined in the latest work of Buben[8]. In this work, the inaccuracy due to the method of development of the exponential, was also eliminated.

The most interesting result is that, at a sufficiently high activation energy, and any given order of reaction m,   critical phenomena are possible if the above-introduced parameter ξ has a value of not less than $(1 + \sqrt{m})^2$.

Stationary Surface Temperature Rise

As we have seen, in strongly exothermal reactions with rates strongly increasing with rising temperature, i.e., with high values of the parameter ξ, one can observe either the lower temperature regime where the surface temperature rise is very small, or the upper temperature regime where the reaction takes place in the diffusional region. Inasmuch as in the lower temperature regime, the temperature of the surface is practically indistinguishable from the temperature of the surrounding medium, the problem of the surface temperature rise is reduced to the calculation of that temperature rise in the upper temperature regime.

Inasmuch as under these conditions the reaction takes place entirely in the diffusional region, the surface temperature attained is entirely independent of the true kinetics of the reaction and is determined solely by the conditions of diffusion and heat exchange.

The problem of the stationary surface temperature rise in the diffusional range was discussed in detail in

the work of Buben[3] carried out under our direction at the
Institute of Chemical Physics.

In the stationary state, the amount of heat
evolved by the reaction at the surface should be equal to
the amount of heat dissipated by heat exchange. We shall
first give a calculation exclusive of the thermal diffusion and Stephan flow. The rate of reaction in the diffusional region can then be represented by formula (II, 8)

$$q = \beta C$$

where C is the reactant concentration in the volume, and
β is the diffusion velocity constant. In order to find
the amount of heat evolved on unit surface per unit time,
it is necessary to multiply that magnitude by the heat
effect of the reaction Q. Let the heat transfer from the
surface take place only to the reacting gas; heat transfer
by radiation or directly through the solid body will be
considered inexistent. Then, the amount of heat removed
from the surface by heat transfer according to the definition (I, 16) of the heat exchange coefficient, will be

$$\alpha(T_1 - T_0)$$

where α is the heat exchange coefficient, T_1 the surface temperature, and T_0 the temperature in the volume.
Consequently, in the stationary state, one should have
the equality

$$Q\beta C = \alpha(T_1 - T_0) \qquad\qquad (IX, 19)$$

The stationary temperature rise of the surface will be expressed as

$$T_1 - T_0 = \frac{\beta}{\alpha} QC \qquad\qquad (IX, 20)$$

The diffusion velocity constant β and the heat exchange coefficient α can be expressed by the Nusselt criterion which (in the presence of convection) can have different values for diffusion and for heat transfer; in the first case it depends on the diffusional, and in the second case on the thermal Prandtl criterion.

Let us designate the diffusional Nusselt criterion by Nu_D, and the thermal by Nu_λ. The diffusion velocity constant will then be expressed by

$$\beta = \frac{Nu_D D}{d}$$

and the heat exchange coefficient by

$$\alpha = \frac{Nu_\lambda \lambda}{d}$$

where D is the diffusion coefficient of the substance, the diffusion of which limits the overall process, λ is the heat conductivity of the gaseous mixture, and d is the determining dimension of the system. Substituting these expressions in (IX, 20), we get

$$T_1 - T_0 = \frac{Nu_D}{Nu_\lambda} \frac{D}{\lambda} QC \qquad\qquad (IX, 21)$$

or, bearing in mind that $\lambda = c\rho a$, where a is the temperature conductivity coefficient, c the heat capacity, and ρ the density:

$$T_1 - T_0 = \frac{Nu_D}{Nu_\lambda} \frac{D}{a} \frac{QC}{c\rho} \qquad\qquad (IX, 22)$$

The magnitude $\frac{QC}{c\rho}$ is nothing but the maximum theoretical temperature rise which would be attained in an adiabatic course of the reaction without any loss of heat. We shall designate the maximum theoretical temperature rise attained

in an adiabatic course of the reaction by $T*$. Having
disregarded the Stephan flow in the calculation of the
true surface temperature, we shall, in the calculation of
the theoretical temperature, disregard the change of the
number of moles in the reaction and the difference of the
heat capacities of the original reactants and the products;
we then have

$$T* - T_0 = \frac{QC}{c\rho} \qquad (IX, 23)$$

and formula (IX, 22) can be written in the form

$$T_1 - T_0 = \frac{Nu_D}{Nu_\lambda} \frac{D}{a} (T* - T_0) \qquad (IX, 24)$$

If all the gases forming the mixture have closely the
same molecular weights, the diffusion coefficient of the
reacting substance will be close to the temperature conductivity coefficient of the mixture. We shall consider
the simples case where these magnitudes are equal,

$$D = a$$

In this case, the values of the thermal and the diffusional Prandtl criterion, and consequently also of the thermal
and the diffusional Nusselt criterion, will be equal, and
formula (IX, 24) will give

$$T_1 - T_0 = T* - T_0 \qquad (IX, 25)$$

Thus, in the simplest case, when the diffusion coefficient
is equal to the temperature conductivity coefficient, the
surface temperature in the upper thermal regime is equal
to the theoretical temperature, i.e., to the temperature
which would be attained if the reaction were carried out
under adiabatic conditions without any heat exchange.

This fact was frequently stressed with respect

to different concrete processes, e.g., the catalytic oxidation of ammonia[4], but it was usually considered as a
direct consequence of thermodynamics, i.e., of the law of
conservation of energy. Actually this is not so. Thermodynamics has nothing to say about the stationary temperature of the surface which is attained under conditions not
in the least adiabatic. The equality of the stationary
surface temperature and the theoretical temperature T*
holds only in the particular instance of equality of the
coefficients of diffusion and of temperature conductivity.
Only if these coefficients are equal, do the equations of
temperature conductance and of diffusion become identical,
which, the similitude of the boundary conditions, leads
to a similitude of the temperature and concentration fields.
The equality of the surface temperature and the theoretical temperature is a consequence of that similitude of
the temperature and the concentration fields, and not at
all of the law of conservation of energy.

In the general case, when the coefficient of
diffusion and of temperature conductivity are not equal,
the stationary surface temperature rise is given by
formula (IX, 24).

In the absence of convection, when the value of
the Nusselt criterion depends only on the geometric configuration of the system, the thermal and the diffusion
Nusselt criterion are equal, and formula (IX, 24) gives

$$T_1 - T_0 = \frac{D}{a} (T* - T_0)$$ \hfill (IX, 26)

In all real instances, convection is present, and the
Nusselt criterion depends on the Reynolds and Prandtl criteria, which dependence can be expressed by formula (I, 41)

$$Nu = k\, Re^m Pr^n$$

Within the limits of accuracy of that formula, we can

consider the magnitudes k, m, and n, to be the same
for heat transfer and for diffusion; this means that we
consider formula (I, 41) with constant k, m, and n,
valid over the whole range of values which the Prandtl criterion can take for both heat transfer and diffusion.

The value of the Reynolds criterion will be the
same for both the thermal and the diffusional Nusselt criterion. In this way, we have

$$Nu_D = k \; Re^m Pr_D^n = k \; Re^m \; (\tfrac{v}{D})^n$$

$$Nu_\lambda = k \; Re^m Pr_\lambda^n = k \; Re^m \; (\tfrac{v}{a})^n$$

and the ratio of these two criteria will be expressed by

$$\frac{Nu_D}{Nu_\lambda} = (\tfrac{a}{D})^n \qquad\qquad (IX, 27)$$

On substitution in formula (IX, 24), we get

$$T_1 - T_0 = (\tfrac{D}{a})^{1-n}(T* - T_0) \qquad (IX, 28)$$

where n is the exponent of the Nusselt criterion in formula (I, 41). Its value depends on the hydrodynamic conditions; in most cases, as we have seen in chapter I, it
is close to 1/3. In all real cases encountered in practice,
the stationary surface temperature rise should be calculated with the convection taken into account.

As can be seen from formula (IX, 28), at $D \neq a$
the temperature of the surface differs from the theoretical temperature which can be either higher or lower. If
the diffusion coefficients of the rate-determining substance,
i.e., of the substance with the lowest value of

$$\frac{\beta_1 c_1}{v_1}$$

is smaller than the temperature conductivity coefficient
of the mixture, the surface temperature will be lower than
the theoretical temperature. If, on the contrary, $D > a$,
the stationary surface temperature in the upper thermal
regime will be higher than the theoretical temperature
corresponding to an adiabatic course of the reaction.

This last conclusion was confirmed experimentally
in Buben's study of the catalytic oxidation of hydrogen on
platinum in mixtures deficient in hydrogen.

Correction for Thermal Diffusion and Diffusion Thermoeffect

Formula (IX, 28) is valid if one may disregard,
first, the external heat exchange, and secondly, the
corrections for thermal diffusion and Stephan flow. By
external heat exchange we mean all forms of heat removal
from the surface besides the direct heat transfer to the
reacting gas, e.g., loss of heat through radiation or
cession of heat by the solid surface through direct contact to other solid bodies (walls of the vessel, ends of
the wire on which the reaction takes place, etc.). The
correction for external heat exchange is easy to introduce
in each concrete case, by adding the corresponding term in
the right-hand member of the equality (IX, 19). Obviously,
this correction depends entirely on the concrete conditions
of the experiment, and warrants no general treatment. In
contrast, the corrections for thermal diffusion and Stephan
flow can be discussed in general form.

It is legitimate to disregard thermal diffusion
in cases where the molecular weights of all the components
of the mixture are close to each other. However, it is
entirely inadmissible in such processes as catalytic oxidation of hydrogen. It is legitimate to disregard the
Stephan flow in cases where the reacting mixture is strongly diluted by inert gases, but not in the presence of high
concentrations of the reactants. The problem of the

calculation of the stationary temperature rise of the surface with the inclusion of thermal diffusion and Stephan
flow has been discussed in detail in the work of Buben[3].
The previous results of Ackerman[5] are erroneous, as this
author has considered only the effect of the Stephan flow
on diffusion and not its effect on heat transfer, whereas
actually the Stephan flow (as any other convective flow)
transfers both matter and heat.

Let us consider first the correction for thermal
diffusion and diffusion thermoeffect. In the presence of
these processes, we must use, for the heat and the diffusion
flow, instead of the usual laws of Fick and Fourier, the
expressions (IV, 9) and (IV, 13).

In the approximation corresponding to the method
of the boundary layer, we shall put

$$(\text{grad } p_1)_n = \frac{p_1^o}{\delta_D}$$

$$(\text{grad } T)_n = -\frac{T_1 - T_o}{\delta_\lambda}$$

where δ_D and δ_λ are the equivalent film thicknesses,
respectively, for the processes of diffusion and of heat
transfer; p_1^o is the partial pressure of the rate-determining gas at a distance from the surface; T_1 is the
temperature of the surface, and T_o the temperature at a
distance from the surface. Formulas (IV, 9) and (IV, 13)
give

$$q_D = \frac{D}{RT}\left(\frac{p_1^o}{\delta_D} - \frac{P}{T}k_T\frac{T_1 - T_o}{\delta_\lambda}\right) \qquad (IX, 29)$$

$$q_\lambda = \lambda\frac{T_1 - T_o}{\delta_\lambda} - JDk_T\frac{P^2}{p_1 p_2}\frac{p_1^o}{\delta_D} \qquad (IX, 30)$$

The velocity of the mass flow v perpendicular to the surface is at this point taken to be zero, as the correction for the Stephan flow is to be considered separately. The thicknesses δ_D and δ_λ of the equivalent film for diffusion and for heat transfer can be different owing to the different values of the Nusselt criterion.

In the stationary state, all the heat evolved by the reaction at the surface, is carried away by heat exchange; consequently, the condition of the stationary thermal state of the surface has the form

$$q_\lambda = - Qq_D \qquad\qquad (IX, 31)$$

where Q is the heat of the reaction. Substituting (IX, 29) and (IX, 30) in (IX, 31) we get

$$\frac{QD}{RT}\frac{p_1^o}{\delta_D} - \frac{QDP}{RT^2}k_T\frac{T_1 - T_o}{\delta_\lambda} = \lambda\frac{T_1 - T_o}{\delta_\lambda} - JDk_T\frac{P^2}{p_1p_2}\frac{p_1^o}{\delta_D}$$

Solving this equation for $T_1 - T_o$ we get

$$T_1 - T_o = \frac{\dfrac{QD}{RT}\dfrac{p_1^o}{\delta_D} + JDk_T\dfrac{P^2}{p_1p_2}\dfrac{p_1^o}{\delta_D}}{\dfrac{\lambda}{\delta_\lambda} + \dfrac{QDP}{RT^2}\dfrac{k_T}{\delta_\lambda}} \qquad\qquad (IX, 32)$$

This formula (IX, 32) is the solution of our problem, as it gives the stationary temperature rise of the surface with the inclusion of thermal diffusion and diffusion thermoeffect. It is easily noted that the temperature rise does not depend on the absolute values of δ_D and δ_λ, only on their ratio. Multiplying and dividing the numerator in (IX, 32) by δ_D, and the denominator by δ_λ, we get

$$T_1 - T_0 = \frac{\delta_\lambda}{\delta_D} \cdot \frac{QD \frac{p_1^o}{RT} + JDk_T \frac{P^2}{p_1 p_2} p_1^o}{\lambda + \frac{QDP}{RT^2} k_T} \qquad (IX, 33)$$

According to formula (I, 28) the thickness of the equivalent film is:

$$\delta = \frac{d}{Nu}$$

and consequently

$$\frac{\delta_\lambda}{\delta_D} = \frac{Nu_D}{Nu_\lambda}$$

where Nu_D and Nu_λ are the diffusional and the thermal values of the Nusselt criterion.

In the absence of convection $Nu_D = Nu_\lambda$ and

$$\frac{\delta_\lambda}{\delta_D} = 1$$

In the presence of convection

$$Nu_D = k \, Re^m \left(\frac{\upsilon}{D}\right)^n$$

$$Nu_\lambda = k \, Re^m \left(\frac{\upsilon}{a}\right)^n$$

whence

$$\frac{\delta_\lambda}{\delta_D} = \frac{Nu_D}{Nu_\lambda} = \left(\frac{a}{D}\right)^n \qquad (IX, 34)$$

Bearing in mind, besides that $\frac{p_1}{RT} = C$, where C is the concentration in the gas phase of the substance the diffusion of which limits the process, we get finally

$$T_1 - T_0 = \left(\frac{a}{D}\right)^n \cdot \frac{QDC + JDk_T \dfrac{p^2}{p_1 p_2} p_1^o}{\lambda + \dfrac{QDP}{RT^2} k_T} \qquad (IX,\ 35)$$

For the magnitudes, T, p_1, p_2, and the physical constants in this formula one should take the mean values for the boundary layer.

We introduce the theoretical temperature $T*$ defined by the relation (IX, 23)

$$T* - T_0 = \frac{QC}{Cp}$$

Formula (IX, 35), on account of $\lambda = c\rho a$, will then take the form

$$T_1 - T_0 = \left(\frac{D}{a}\right)^{1-n}(T* - T_0)\frac{1 + J\dfrac{k_T}{QC}\dfrac{p^2}{p_1 p_2} p_1^o}{1 + \left(\dfrac{D}{a}\right)\dfrac{T* - T}{CRT^2}Pk_T} \qquad (IX,\ 36)$$

This expression differs from formula (IX, 28) by the two corrective terms in the numerator and the denominator. The first allows for diffusion thermoeffect, the second for thermal diffusion.

Bearing in mind that $Jp_1^o = RTC$, where R is expressed in thermal units, and introducing the mole fraction of the diffusing substance in the gaseous mixture

$$x = \frac{C}{\Sigma C} = \frac{CRT}{P}$$

we bring (IX, 36) to the form

$$T_1 - T_0 = \left(\frac{D}{a}\right)^{1-n} (T*-T_0) \frac{1 + \frac{RT}{Q} k_T \cdot \frac{P^2}{P_1 P_2}}{1 + \left(\frac{D}{a}\right) \frac{T* - T_0}{x} \frac{k_T}{T}} \qquad (IX, 37)$$

where R is expressed in thermal units. Formula (IX, 37) represents the final formula for the computation of the stationary temperature rise at the surface with the inclusion of thermal diffusion and diffusion thermoeffect, in the presence of convection.

For mixtures strongly diluted by an inert gas, this formula can be simplified. In this case we can put

$$p_2 \approx P$$

$$\frac{p_1}{P} = x$$

and make use of the relation (IV, 3)

$$k_T = bx$$

where b is a constant. After that, we get from (IX, 37)

$$T_1 - T_0 = \left(\frac{D}{a}\right)^{1-n} (T* - T_0) \frac{1 + \frac{RTb}{Q}}{1 + \frac{D}{a} \frac{T* - T_0}{T} b} \qquad (IX, 38)$$

For reactions with a large heat effect, only the corrective term in the numerator, i.e., only the correction for thermal diffusion, is essential. The correction for diffusion thermoeffect is negligibly small for such reactions.

In this case, if $k_T > 0$, i.e., if the process is limited by the diffusion of the heavier gas, thermal diffusion will lower the surface temperature, and vice versa. These conclusions of the theory have been confirmed experimentally by Buben's[3] study of the thermal regime of the surface in the catalytic oxidation of hydrogen on platinum.

In the preceding calculation we did not take into account the temperature dependence of the physical constants. That is why the formulas obtained include, in addition to the temperature T_1 and T_0 which have a definite meaning, also a mean temperature T for the numerical value of which one usually takes the arithmetic mean of the temperatures T_1 and T_0.

This method of calculation, and the use of the arithmetic mean temperature, are adequate only for small temperature differences. Accurate calculation in the case of a large temperature rise calls for the inclusion of the temperature dependence of k_T (or b) and of other physical constants, which results in quite complex expressions.

Correction for the Stephan Flow

In calculating the effect of the Stephan flow on the stationary surface temperature, we can use directly the expression for the flow of material (III, 25). If, as we have been assuming all the time, heat is transferred from the surface only through direct contact with the reacting gas, it follows from the law of conservation of energy

$$- \lambda (\mathrm{grad}\ T)_n + \sum_i q_i I_i = 0 \qquad (IX, 39)$$

where w_i is the diffusion flow of the ith component of the mixture, and I_i its heat content, figured from the common zero point of energy, i.e., including both physical heat and the chemical energy. The expression (IX, 39) is entirely analagous to the formula (IX, 31) but is more

convenient for the discussion of the Stephan flow, as it
spares the calculation of the total heat flow including t
the amount of heat transferred by the Stephan flow. With
the aid of the condition of stoichiometry (III, 1), we
can transform (IX, 39) into

$$- \lambda(\text{grad } T)_n + \frac{q_1}{v_1} \sum_i v_i I_i = 0 \qquad (\text{IX, 40})$$

where v_i are the stoichiometric coefficients, negative
for the original reactants and positive for the final
products; the subscript 1 refers to the substance the
diffusion of which limits the process, i.e., the substance
for which the magnitude $\frac{\beta_1 C_1}{v_1}$ has the smallest value. We
pass now to the individual zero point of energy, thus
separating the chemical energy from the physical heat. To
that end, we represent the heat content in the form

$$I_1 = I_1^O + c_1 (T - T^O)$$

where T^O is an arbitrary temperature, and I^O is the
heat content at that temperature. The heat capacity c
is calculated per mole, as is the heat content. Formula
(IX, 40) then becomes

$$- \lambda(\text{grad } T)_n + q_1[Q^O + (T - T^O) \frac{1}{v_1} \cdot \sum_i v_i c_i] = 0 \qquad (\text{IX, 41})$$

where

$$Q^O = \sum_i v_i I_i^O$$

is the heat of the reaction at the temperature T^O under
constant pressure, per one mole of the substance designated
by the subscript 1. Formula (IX, 41) can be written in
the form

$$- Q^{o}q_1 = - \lambda (grad\ T)_n + q_1 \frac{T - T^{o}}{v_1} \sum_i v_1 c_1$$

Combining with (IX, 31) we conclude that the amount of heat transferred by the Stephan flow is expressed by

$$q_1 (T - T^{o}) \cdot \frac{1}{v_1} \sum_i v_1 c_1$$

where T^{o} is the temperature to which the given heat of reaction refers. Thus, allowance for the amount of heat transferred by the Stephan flow is identical with an allowance for the temperature dependence of the heat of the reaction.

Integration of equation (IX, 41) over the boundary layer gives

$$q_1 = \frac{\lambda}{\delta_\lambda} \frac{v_1}{\sum_1 v_1 c_1} \ln \frac{Q^{o} + \frac{1}{v_1} \sum_1 v_1 c_1 (T_1 - T^{o})}{Q^{o} + \frac{1}{v_1} \sum_1 v_1 c_1 (T_0 - T^{o})} \qquad \text{(IX, 42)}$$

where T_1 is the surface temperature and T_0 the temperature in the volume. Considering that according to Kirchhoff's law, the temperature dependence of the heat of reaction, in our notation, is expressed by

$$\frac{dQ}{dT} = \frac{1}{v_1} \sum_1 v_1 c_1$$

we can represent formula (IX, 42) in the form

$$q_1 = \frac{\lambda}{\delta_\lambda} \frac{v_1}{\sum_1 v_1 c_1} \ln \frac{Q(T_1)}{Q(T_0)} \qquad \text{(IX, 43)}$$

where $Q(T_0)$ and $Q(T_1)$ are the values of the heat of the reaction at the temperatures T_0 and T_1. This formula

(IX, 43) does not contain the arbitrary temperature T^o and, consequently, the choice of that temperature does not alter the result of the calculation. In order to calculate the surface temperature T_1, it is necessary to substitute in formula (III, 25)

$$q_1 = - \frac{P}{\gamma RT} \frac{D}{\delta_D} \ln\left(1 - \gamma \frac{p_1^o}{P}\right) = - \frac{P}{\gamma RT} \frac{D}{\delta_D} \ln(1 - \gamma x_1)$$

and solve the equation obtained for T_1. We put in (IX, 42) $T^o = T_o$ and substitute for q_1 its expression (III, 25); we then get

$$- \frac{D}{\delta_D} \frac{P}{\gamma RT} \ln(1 - \gamma x_1) = \frac{\lambda}{\delta_\lambda} \frac{v_1}{\sum_1 v_1 c_1} \ln\left[1 + \frac{\sum_1 v_1 c_1}{v_1 Q(T_o)}(T_1 - T_o)\right] \quad (IX, 44)$$

We introduce the designations

$$\frac{\sum_1 v_1 c_1}{v_1 Q(T_o)} = \sigma \qquad (IX, 45)$$

$$\frac{\sum_1 v_1 c_1}{v_1 \gamma \bar{c}} = \frac{\sum v_1 c_1}{D \bar{c} \sum \frac{v_1}{D_1}} = \tau \qquad (IX, 46)$$

where $\bar{c}$ is the mean heat capacity of the gaseous mixture per one mole, so that

$$\lambda = \bar{c} a \Sigma C$$

In the notation adopted, (IX, 44) takes the form

$$\ln[1 + \sigma(T_1 - T_o)] = - \left(\frac{D}{a}\right)^{1-n} \tau \ln(1 - \gamma x_1) \qquad (IX, 47)$$

or

$$1 + \sigma(T_1 - T_0) = (1 - \gamma x_1)^{-\left(\frac{D}{a}\right)^{1-n}\tau} \qquad \text{(IX, 48)}$$

Here x_1 is the mole fraction, in the gaseous mixture, of the substance the diffusion of which limits the process.

In all the formulas of this chapter, heat capacity is understood as the mean heat capacity in the temperature range between the unknown surface temperature T_1 and the temperature T_0 in the gas volume.

The magnitude γ is defined by the formula

$$\gamma = \frac{D}{v_1} \Sigma \frac{v_i}{D_i} \qquad \text{(III, 14)}$$

Formula (IX, 48) serves to determine the stationary surface temperature rise with the inclusion of the Stephan flow. It is easily seen that when the Stephan flow is directed away from the surface, it will lower the surface temperature. If, on the contrary, the Stephan flow points to the surface, the temperature can be higher than the theoretical temperature.

This conclusion is valid for catalytic reactions involving only gaseous substances. If a solid substance is formed or consumed in the reaction, it is necessary to take into account the amount of heat remaining in the solid at its formation or evolved from it at its consumption. In this case we have, instead of (IX, 40)

$$- \lambda(\text{grad } T)_n + \frac{q_1}{v_1} \Sigma_i v_i I_i = - q_1 \frac{v_s I_s}{v_1} \qquad \text{(IX, 49)}$$

where v_s is the stoichiometric coefficient of the solid and I_s its heat content.

We consider a quasistationary state when the

temperature of the solid is equal to the surface temperature T_1, and remains constant. Then

$$I_s = c_s (T_1 - T_o)$$

By a calculation analogous to the foregoing, we get instead of (IX, 48) the following equations for the determination of $T_1 - T_o$

$$1 + \frac{\sigma (T_1 - T_o)}{1 + \frac{v_s c_s}{v_1 Q(T_o)} (T_1 - T_o)} = (1 - \gamma x_1)^{- \left(\frac{D}{a}\right)^{1-n} \tau} \qquad \text{(IX, 50)}$$

$$T_1 - T_o = \frac{[(1 - \gamma x_1)^{- \left(\frac{D}{a}\right)^{1-n} \tau} - 1] v_1 Q(T_o)}{\sum_1 v_1 c_1 + v_s c_s [1 - (1 - \gamma x_1)^{- \left(\frac{D}{a}\right)^{1-n} \tau}]} \qquad \text{(IX, 50a)}$$

In this case, it is no more necessary to take into account
the stoichiometric coefficient of the solid in the calculation of γ, σ, and τ.

Let us consider specific concrete cases of application of these formulas. The simplest case is that of a
mixture strongly diluted by inert gases or by one of the
reacting components. In this case

$$x_1 \ll 1$$

and, as γ is always of the order of unity,

$$\gamma x_1 \ll 1$$

Developing the right-hand member of equation (IX, 48) into a Newtonian binomial, we get

$$1 + \sigma(T_1 - T_0) = 1 + \left(\frac{D}{a}\right)^{1-n} \tau\gamma x_1$$

$$+ \frac{1}{2}\left(\frac{D}{a}\right)^{1-n}\tau\left[\left(\frac{D}{a}\right)^{1-n}\tau + 1\right]\gamma^2 x_1^2 + \cdots \qquad (IX, 51)$$

whence

$$T_1 - T_0 = \left(\frac{D}{a}\right)^{1-n}\frac{\tau\gamma x_1}{\sigma} + \frac{1}{2}\left(\frac{D}{a}\right)^{1-n}\tau\left[\left(\frac{D}{a}\right)^{1-n}\tau + 1\right]\frac{\gamma^2 x_1^2}{\sigma} + \cdot$$

Substituting for γ, σ, and τ, their values (III, 14), (IX, 45), and (IX, 46), we get, considering (IX, 23),

$$\frac{\tau\gamma x_1}{\sigma} = \frac{\Sigma v_1 c_1}{v_1\bar{c}}\frac{x_1 v_1 Q(T_0)}{\Sigma v_1 c_1} = \frac{Q(T_0)x_1}{\bar{c}} = \frac{Q(T_0)C_1}{\bar{c}\Sigma C} = T* - T_0$$

If the gaseous mixture is strongly diluted by an inert gas or an excess of one of the reacting gases, its heat capacity will be determined mainly by the heat capacity of the latter. In this case it is entirely legitimate to disregard the difference between the heat capacities of the original reactants and the reaction products and to consider that the theoretical temperature corresponding to an adiabatic course of the reaction, $T*$, is defined by formula (IX, 23).

Thus, formula (IX, 51) is reduced to

$$T_1 - T_0 = \left(\frac{D}{a}\right)^{1-n}(T* - T_0)\left\{1 + \frac{1}{2}\left[\left(\frac{D}{a}\right)^{1-n}\tau + 1\right]\gamma x_1\right\}(IX, 52)$$

In a first approximation, at small γx_1, one can disregard the second term in { }, and then formula (IX, 52) will coincide with (IX, 28). This bears our our previous contention that, in dilute mixtures, one can disregard the effect of the Stephan flow. In a second approximation, formula (IX, 52) gives a corrective term which allows for the effect of the Stephan flow on the surface temperature at small γx_1.

If a solid body is formed or consumed in the reaction, formula (IX, 50) will give at $\gamma x_1 \ll 1$

$$T_1 - T_0 = \left(\frac{D}{a}\right)^{1-n}(T^* - T_0) \frac{1 + \frac{1}{2}\left[\left(\frac{D}{a}\right)^{1-n}\tau + 1\right]\gamma x_1}{1 - \frac{v_s c_s}{v_1 \bar{c}}\left(\frac{D}{a}\right)^{1-n}x_1} \quad (IX, 52a)$$

Thus, in dilute mixtures, formation or consumption of a solid phase affects only the corrective term containing x_1.

We shall consider another simple case: let the diffusion coefficients of all substances forming the mixture be the same and equal to the temperature conductivity coefficient of the mixture, and let the heat capacities of all reactants and products be equal.

In this case, formulas (III, 14), (IX, 45), and (IX, 46) give

$$\gamma = \frac{\Sigma v_1}{v_1}$$

$$\tau = 1$$

$$\sigma = \frac{\Sigma v_1}{v_1}\frac{\bar{c}}{Q(T_0)}$$

We take at once the general case when the reaction involves not only gaseous substances but also a solid surface at which the reaction takes place [formula (IX, 50)]. The exponent in that formula will be equal to unity and formula (IX, 50) will become

$$1 + \frac{\Sigma v_1}{v_1} \frac{\bar{c}}{Q(T_o)} \frac{T_1 - T_o}{1 + \frac{v_s c_s}{v_1 Q(T_1)} (T_1 - T_o)} = \frac{1}{1 - \frac{\Sigma v_1}{v_1} x_1} \qquad (IX, 53)$$

whence

$$T_1 - T_o = \frac{x_1}{\frac{\bar{c}}{Q(T_o)} - \frac{\Sigma v_1}{v_1} \frac{\bar{c}}{Q(T_o)} x_1 - \frac{v_s c_s}{v_1 Q(T_o)} x_1}$$

or

$$T_1 - T_o = \frac{\frac{Q(T_o)}{\bar{c}} x_1}{1 - \frac{x_1}{v_1} \left(\Sigma v_1 + \frac{v_s c_s}{\bar{c}} \right)} \qquad (IX, 54)$$

The magnitude

$$\frac{Q(T_o) x_1}{\bar{c}}$$

coincides with the definition of $T^* - T_o$ according to formula (IX, 23). But, in this case, the temperature T^*, thus defined, is not any more equal to the theoretical temperature for an adiabatic course of the reaction.

Actually, at high surface temperatures, particularly in heterogeneous combustion processes, there usually arise additional reactions in the volume. In the first place, at high surface temperatures, dissociation of the combustion products becomes significant, as a result of which the final products are formed only after complete cooling. On the other hand, the temperature of the gas in the neighborhood of the surface becomes high enough to give

rise to various homogeneous reactions. All these circumstances evidently complicate the problem of the thermal
regime of the surface. The foregoing calculations are
valid only for the simplest case where homogeneous reactions are absent. This situation can mostly be realized
only if the surface temperature is kept sufficiently low
through dilution of the reacting mixture by inert gases.
But then all the corrections discussed in the foregoing,
except the correction for thermodiffusion, become negligible.

Thus, the practically most important formulas
for the calculation of the thermal regime of the surface
are (IX, 28) and, for the case of a large difference in
the molecular weights of the gases, (IX, 38).

Experimental Data

The thermal regime of exothermal heterogeneous
reactions was studied, in the instance of the catalytic
oxidation of hydrogen and other fuels on platinum wires,
by Davies[7], and in greater detail by Buben[3].

Figures 25 - 28 show some of the experimental
data of Buben.

In Figures 25 and 26, the ordinates represent
the temperature of the wire as a function of the intensity
of the heating current. Figure 25 refers to a mixture of
hydrogen and air, and Figure 26 to a mixture of ammonia and
air. The lower branch of each curve refers to the lower
thermal regime. Here, the temperature rise of the surface
does not depend on the fuel content in the mixture.

This temperature rise is due entirely to the
heating of the wire by the current. If the heating current
intensity is increased and thus the wire temperature raised,
a discontinuous upward jump of the temperature (marked by
arrows on the figures) will occur when a definite temperature is attained (about 100°C for hydrogen, and about 200°C
for ammonia). The process passes onto the upper branch of

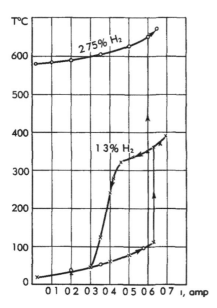

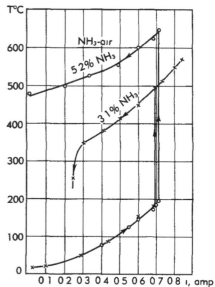

FIGURE 25. SURFACE TEMPERATURE AS
A FUNCTION OF THE INTENSITY OF THE
HEATING CURRENT IN THE CATALYTIC
OXIDATION OF HYDROGEN ON PLATINUM
(BUBEN)

Ordinates are surface temperatures
in degrees centigrade. Abscissas
are intensities of the heating
current in amperes. The circles
represent the results of experiments
with a mixture of hydrogen and air,
containing 2.75% hydrogen; the
crosses refer to a mixture with 1.3%
hydrogen. The arrow points to the
beginning of ignition.

FIGURE 26. SURFACE TEMPERATURE AS A
FUNCTION OF THE INTENSITY OF THE
HEATING CURRENT IN THE CATALYTIC
OXIDATION OF AMMONIA ON PLATINUM
(BUBEN)

(Ordinates and abscissas, as in
Figure 25) The circles refer to
results of experiments with a mixture of ammonia and air containing
5.2% ammonia; the crosses refer to
a mixture with 3.1% ammonia.

the curve, corresponding to the upper temperature regime.
In this range, the temperature rise at the surface depends on the fuel content in the mixture; the figures show
two curves, each referring to a different fuel content.
If, after ignition of the surface, the heating current intensity is made to decrease, the upper temperature regime
will persist at a substantially lower current intensity
than was needed for the ignition. At a low fuel content
in the mixture, decrease of the heating current intensity
will finally cause extinction of the surface, i.e., sharp
passage onto the lower branch of the curve. Extinction
takes place at a considerably higher temperature of the
surface than the ignition. For mixtures of hydrogen with
air, that temperature is about 300° and for mixtures of
ammonia with air, about 350°C.

At high fuel contents in the mixture, even complete shutting off of the current will not result in an
extinction of the surface. In such cases, it is possible
to measure directly the stationary temperature rise of the
surface as a result of the reaction in the absence of a
heating current. In poor mixtures, where elimination of
the heating current results in extinction of the surface,
the stationary temperature rise of the surface can be determined from the difference of temperatures on the upper
and the lower branches of the curve.

In Figures 27 and 28, the stationary temperature
rise of the surface is represented as a function of the
fuel content of the mixture. Figure 27 refers to mixtures
deficient in hydrogen, in which the process is limited by
the diffusion of hydrogen. Here, the abscissa represents
the percent content of hydrogen in the mixture. In each
figure, the curve a represents the stationary temperature
rise, calculated with inclusion of the effect of the
Stephan flow (which, on account of the low content of the
diffusing gas in the mixture, is slight) but without corrections for thermal diffusion and radiation. The curve

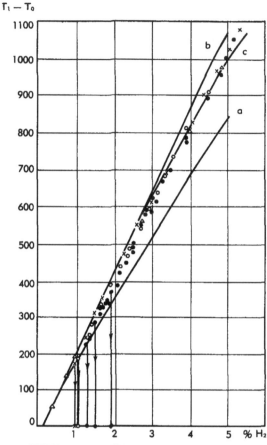

$T_1 - T_0$

FIGURE 27.   STATIONARY TEMPERATURE RISE OF THE
SURFACE IN MIXTURES DEFICIENT IN HYDROGEN
(BUBEN)

Ordinate represents the surface temperature rise
$T_1 - T_0$. Abscissa gives the percent H_2.   Curve a
was drawn with inclusion of the Stephan flow only;
curve b, with inclusion of thermal diffusion;
curve c, with inclusion of loss of heat by radiation.   The different kinds of points correspond to
different velocities of the gas flow.

b   represents the stationary temperature rise of the surface, calculated with the thermal diffusion included.   In
mixtures where the process is limited by diffusion of the
lighter gas, thermal diffusion enhances the temperature
rise of the surface, as in this case the thermal diffusion
flow is in the same direction as the ordinary diffusion
flow, and therefore the total velocity of diffusion is

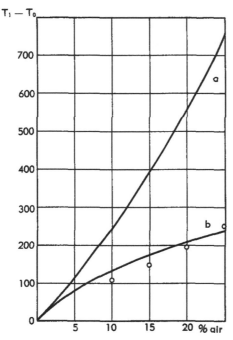

FIGURE 28.   STATIONARY TEMPERATURE RISE
OF THE SURFACE IN MIXTURES WITH EXCESS
HYDROGEN (BUBEN)

Ordinates represent the surface temperature
rise T_1 - T_0. Abscissas give the percent
content of air in the mixture. Other
designations are the same as in Figure 27.

greater than without thermal diffusion. In contrast, in
mixtures where the process is limited by diffusion of the
heavier gas (in this case, in mixtures with excess hydrogen), the direction of the thermal diffusion flow is
opposed to the ordinary diffusion flow and therefore causes
a lowering of the surface temperature.

Curve c in Figure 27 was drawn with inclusion
of loss of heat by radiation. In mixtures with excess hydrogen, this correction can be disregarded, the temperatures involved being low enough for the intensity of
emission to be negligible. The experimentally measured
surface temperatures are represented by points. The
different kinds of points refer to different velocities of
the gas flow. The stationary surface temperature rise is,

evidently, practically independent of the gas flow velocity, and is in fair agreement with the calculation.

The influence of thermal diffusion on the surface temperature rise, in mixtures of hydrogen and air, turns out to be very strong; this is due to the great difference of the molecular weights of the components of the mixture. In other mixtures of combustible gases, this influence will be very much weaker.

Earlier, experimental data of the surface temperature rise of a platinum wire were obtained by Davies[7] who also came to the qualitative conclusion that the rate of the process is limited by diffusion.

Thermal Regime of a Layer or Channel

We have considered the thermal regime of a uniformly accessible surface, all points of which are under the same conditions of diffusion and heat exchange. Correspondingly, the reactant concentration and the temperature in the volume were considered to be the same everywhere.

Of great practical importance is the case where the flow of reactants passes through a layer or channel of a length sufficient for the concentration and temperature to change considerably. The picture of the phenomena is not changed in principle; the same stationary thermal regimes and critical conditions are preserved. However, a number of complications arise on closer consideration, linked with the longitudinal transfer of heat and with the possibility of propagation of the reaction zone along the layer or the channel.

A complete theory which would take into account the longitudinal heat conduction in a general form, does not exist at the present time. What does exist in the literature is a number of approximate calculations for sundry simple special cases. Thus, the work of Mayers[17] discusses

the thermal regime of a layer of burning coal, and the
work of Todes and Margolis[16], the thermal regime of a
catalyst bed for two simple limiting cases. In the first
instance, that of a "long layer" of catalyst, the authors
simply disregard the longitudinal heat transfer and take
into account only the change of concentration and temperature along the layer resulting from the progress of the
reaction; in the second instance, that of a "short layer",
they consider the longitudinal heat transfer intense
enough for the temperature to be assumed constant over the
whole length of the layer.

We are not going to enter into detailed quantitative calculations and shall limit ourselves to fundamental qualitative considerations. We shall remark first
of all that, in this problem, there can be two forms of
critical conditions of ignition and of extinction. We
shall designate by the term of critical conditions of the
first kind, the condition of ignition and extinction discussed in the foregoing and linked with removal of heat
as a result of heat transfer. If the longitudinal heat
conductivity is large enough to result in an equalization
of the temperature along the layer, critical conditions of
the second kind, linked with removal of heat by the products of the reaction, will become possible. Critical conditions of the second kind can arise also in a homogeneous
exothermal reaction in a flow, under conditions of complete mixing of the ingoing and the reacting mixtures.
The idea of the existence of critical conditions of the
second kind was first suggested by Zeldovich[15] and Zysin,
particularly with respect to homogeneous reactions.

The physical meaning of critical conditions of
the second kind can be illustrated in the following way.
The cause of the appearance of the critical condition of
extinction is the fact that the rate of the reaction cannot increase indefinitely with the temperature. In the
case of critical conditions of the first kind, this increase

is halted by the passage of the reaction into the diffusional region. Under the critical conditions of the second
kind, an analogous result is attained for an entirely different reason. The total amount of reacting gas within a
layer or a channel is limited. With increasing temperature,
evolution of heat increases with increasing rate of reaction only up to the point where the whole stock of reacting gas is fully spent within the time of its stay in the
reaction zone. Further increase of the rate of reaction
can lead to no further increase of the overall heat evolution, unless the velocity of the gas flow is increased.
This can result in critical conditions of ignition and extinction formally entirely analagous to, but basically
different from the critical conditions of the first kind.
By their nature, these critical conditions are akin to
those formulated in the work of Zeldovich[15] for homogeneous
reactions.

Critical conditions of the first kind are intimately linked with the passage of the reaction from the
kinetic into the diffusional region and vice versa. In
contrast, critical conditions of the second kind have
nothing to do with the diffusional region and can be observed within the kinetic region. The stationary surface
temperature jumps at ignition, but the reaction remains
within the kinetic region.

The possibility of critical conditions within
the kinetic region, in a reaction in a layer of catalyst,
was pointed out in the work of Todes and Margoulis[16] in the
instance of the catalytic oxidation of isooctane in a flow.

We had, even earlier, pointed out that possibility in the instance of combustion of a carbon channel.

We shall now consider our mathematical theory
of critical phenomena of the second kind for a very much
idealized scheme, mainly for the purpose of sketching a
qualitative picture.

We shall consider the temperature of the reacting surface to be the same along the whole length of the
layer or channel. We shall designate that temperature by
T and seek to determine it from the equation of the heat
balance

$$QVS(C_0 - C) = \alpha\sigma(T - T_0) \qquad (IX, 55)$$

where Q is the heat of the reaction, V the linear
velocity of the gas flow, S the free cross-section area
of the layer or channel, C_0 and C the concentrations
of the reacting gas at the entrance and exit of the
channel, σ the reacting surface area, α the heat transfer coefficient with the surrounding medium, and T_0 the
temperature of that medium.

We shall consider the simplest case where the
critical conditions lie within the kinetic region and the
kinetics of the reaction is of the first order. The law
of the change of the concentration of the reacting gas
along the layer or channel will then be of the form

$$\ln \frac{C}{C_0} = -\frac{k\sigma}{VS} \qquad (IX, 56)$$

where k is the rate constant, depending on the temperature according to Arrhenius law

$$k = ze^{-E/RT}$$

Combining (IX, 55) and (IX, 56) we get for the
determination of the stationary temperature T the
equation

$$\frac{k\sigma}{VS} = -\ln\left[1 - \frac{\alpha\sigma(T - T_0)}{QVSC_0}\right] \qquad (IX, 57)$$

In the following, taking into account that $E \gg RT$, we

shall avail ourselves of the development of the exponential
which we have used previously in the theory of thermal
ignition

$$k = ze^{-\frac{E}{RT}} \approx ze^{-\frac{E}{RT_o}}e^{\theta} \qquad (IX, 58)$$

where

$$\theta = \frac{E}{RT_o^2}(T - T_o)$$

is the dimensionless temperature. With the aid of the
development (IX, 58) we bring (IX, 57) to the form

$$\frac{ze^{-\frac{E}{RT_o}}}{VS}e^{\theta} = -\ln\left[1 - \frac{RT_o^2}{E}\frac{\alpha\sigma}{QVSC_o}\theta\right]$$

or

$$\zeta e^{\theta} = -\ln(1 - \upsilon\theta) \qquad (IX, 59)$$

where

$$\zeta = \frac{ze^{-\frac{E}{RT_o}}\sigma}{VS}$$

$$\upsilon = \frac{RT_o^2}{E}\frac{\alpha\sigma}{QVSC_o}$$

are dimensionless parameters.

Solving equation (IX, 59) with respect to θ,
we can find the stationary temperature rise as a function
of two dimensionless parameters, ζ and υ.

In order to elucidate the general properties of

the solutions, we write (IX, 59) in the form

$$\zeta = - \ln(1 - \upsilon\theta)e^{-\theta} = f(\theta) \qquad (IX, 60)$$

We plot θ as abscissas, and $f(\theta)$ as ordinates. The point of intersection of the curve with the straight line running parallel to the axis of abscissas at the ordinate ζ gives the desired stationary temperature.

If $f(\theta)$ has neither maxima nor minima, the stationary solution will exist always and will be unique. If $f(\theta)$ does have maxima or minima, several stationary states are possible, and the values of the parameters at which a maximum or a minimum occur will give the critical conditions of transition from one regime to another.

To find such critical condition it is necessary, therefore, to solve the equation

$$\frac{df(\theta)}{d\theta} = 0 \qquad (IX, 61)$$

with respect to θ and to substitute the value of θ in the equation (IX, 59). We then get a relation between the parameters ζ and υ, corresponding to the critical condition.

Equation (IX, 61) is reduced to

$$- (1 - \upsilon\theta) \ln(1 - \upsilon\theta) = \upsilon$$

or

$$- x \ln x = \upsilon \qquad (IX, 62)$$

where

$$x = 1 - \upsilon\theta \qquad (IX, 63)$$

Equation (IX, 62) is easily solved with the aid of Figure 29 where the abscissas are x and the ordinates $x \ln x$.

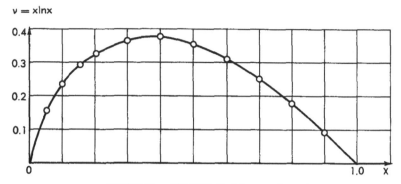

FIGURE 29. SOLUTION OF EQUATION (IX, 62)
Ordinates are $v = -x \ln x$.
Abscissas are x.

It is easily seen that $x \ln x$ has a minimum at $x = \frac{1}{e}$,
and that the minimum value of $x \ln x$ is equal to $-\frac{1}{e}$.
Consequently, equation (IX, 62) has no solution at all at
$v > \frac{1}{e}$, and two different values of x correspond to every
v at $v < \frac{1}{e}$. This means that $f(\theta)$ is a monotonous
function at $v > \frac{1}{e}$, and has two extreme values, one maximum and one minimum, at $v < \frac{1}{e}$. In the first case equation (IX, 59) has always only one solution; in the second
case, three solutions are possible. As was shown above
in a formally analogous case, only the two extreme solutions out of the three possible correspond to stable states.
Thus, there are no critical phenomena at $v > \frac{1}{e}$; the surface temperature changes continuously with the change of
the external conditions. In contrast, at $v < \frac{1}{e}$, there
arise critical conditions of ignition and extinction.

To find these conditions, we proceed as follows.
Prescribing some value of v, we find the two corresponding values of x from the curve in Figure 29, and, according to (IX, 63), two values of θ, the smaller of which
corresponds to the critical condition of ignition and the
larger of which to the condition of extinction. Substituting these values of θ in equation (IX, 59), we find
two values of ζ at which, at the given v, ignition and
extinction will occur.

For practical computations it is convenient to introduce instead of ζ the new parameter

$$\delta = \frac{\zeta}{\upsilon} = \frac{E}{RT_o^2}\frac{Q}{\alpha}C_o z e^{-\frac{E}{RT_o}}$$

while the parameters ζ and υ are both dependent on the velocity V of the gas flow, δ depends only on the temperature, pressure, and composition of the gas, and on the conditions of the heat exchange. This parameter is entirely analogous to that introduced, with the same symbol, in the theory of thermal ignition in homogeneous reactions (see chapter VI). The critical values of δ, calculated in this way for different values of υ, are represented in Figure 30. The upper branch of the curve corresponds to the critical condition of ignition, the lower to the condition of extinction. At υ tending to zero, the critical value of δ tends to $\frac{1}{\upsilon}$. This result has already been obtained in the foregoing.

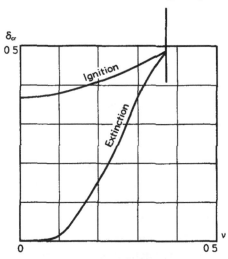

FIGURE 30. CRITICAL CONDITIONS FOR A
            LAYER OR CHANNEL

Ordinates are δ_{cr}. Abscissas are υ.
The upper curve gives the critical condition of ignition, the lower the condition of extinction.

Applications

The question of the thermal regime of a reacting solid surface and of the critical conditions of its ignition and extinction has a practical significance both for the process of combustion of carbon[1,6] and for some strongly exothermal heterogeneous-catalytic processes such as, for example, the contact oxidation of ammonia to nitric acid and of alcohols to aldehydes or ketones. In this process, high surface temperature rise is entirely permissible and in no way harmful; consequently, these processes are always carried out in the upper thermal regime, in the diffusional region, with a considerable temperature difference between the surface and the gas.

In contrast, in such processes as the production of synthetic gasoline from water gas or the catalytic oxidation of hydrocarbons to aldehydes, alcohols, and oxides, high surface temperature rise is unwanted, as the process is apt to go in a wrong direction at too high a temperature. In such cases one must either endeavor to keep the process at the lower thermal regime or to reduce drastically the temperature rise attained in the diffusional region.

To classify heterogeneous exothermal processes by their thermal regime, we can avail ourselves of the dimensionless parameter ξ introduced above [formula (IX, 11)].

Processes for which the value of this parameter is small, are characterized by a small surface temperature rise which varies continuously with the variation of the external conditions. In such processes, one can disregard the temperature difference between the surface and the gas and the temperature distribution over the cross-section of the contact apparatus; to each section of the contact apparatus, one can ascribe one definite temperature, as is usually done in technical calculations.

In contrast, at large values of the parameter ξ, high local temperature differences between the surface and the gas volume become possible.

If such a temperature rise is undesirable, there are several ways to combat it: we can either carry the process into the lower temperature range (into the kinetic region) or endeavor to diminish the temperature rise attained in the upper temperature range (in the diffusional region).

Dilution of the gas mixture by an excess of one of the reactants or with an inert gas results in a decrease of the parameter ξ which is proportional to the concentration C of the reacting substance, and, consequently, in a lowering of the stationary temperature rise attained in the upper temperature range. At a sufficient dilution, the parameter ξ can become so small (for a first-order reaction, less than 4) that the difference between the upper and the lower regime will vanish altogether. That is why, in practice, one endeavors to carry out strongly exothermal processes in strongly diluted mixtures.

Thermal Regime of Contact Apparatus

If the process is carried out in a contact apparatus of the usual type, i.e., in tube-shaped reactor filled with the contact mass, the condition for the possibility of having the process takes its course in the lower temperature range, i.e., in the kinetic region, is identical with the condition for the possibility of a stationary temperature distribution over the reaction of the apparatus in a reaction taking place in the kinetic region. To find this condition, we shall use directly the results exposed in Chapter VI.

According to formula (VI, 16), the stationary temperature distribution in a cylindrical vessel which is the seat of an exothermal chemical reaction, is possible until the critical value

$$\delta_{cr} = 2$$

is reached for the parameter δ defined by the formula

$$\delta = \frac{Q}{\lambda} \frac{E}{RT_o^2} r^2 z e^{-\frac{E}{RT_o}}$$

In this instance, λ represents the effective heat conductivity of the contact mass, with the inclusion of heat transfer from one lump to another both through direct contact and through radiation or across the reacting gas.

The magnitude $ze^{-\frac{E}{RT}}o$ is the rate of reaction per unit volume of the contact mass. Formula (VI, 14) shows that, if one increases the cross-section of the tube (or its radius r), it is necessary, in order to preserve a constant value of δ, to decrease the rate of reaction (for example, by lowering the temperature or decreasing the activity of the contact) inversely proportionally to r^2, i.e., so that the total amount of substance reacting over the whole area of the cross-section of the reactor tube, remains constant. Consequently, the condition of absence of a local temperature rise limits the full productivity of the reaction tube per unit length.

The amount of heat evolved as a result of the reaction per unit length of the reaction tube is

$$Q\pi r^2 z e^{-\frac{E}{RT_o}}$$

According to formulas (VI, 14) and (VI, 16), this magnitude must not be greater than

$$\delta_{cr}\pi \frac{\lambda}{E/RT_o^2}$$

for the process to be able to take its course in the lower

temperature range. The magnitude E/RT_0^2 can be termed
the exponential temperature coefficient of the rate of
reaction.

For the process to take place in the lower temperature range the amount of heat evolved per unit length of
the circular-section reaction tube must not be greater than
2π times the ratio of the effective heat conductivity of
the contact mass and the exponential temperature coefficient of the rate of reaction.

It is possible to carry the process into the
lower temperature range by reducing the rate of reaction
per unit volume of the contact mass, e.g., by lowering the
activity of the contact mass or by diluting it with an
inert filler. If the process is to take place in the lower temperature range, one cannot use an increase of the
cross-section of the reaction tube as a means of increasing
the total space-time yield of the apparatus, but it is
necessary either to increase the number of the reaction
tubes or their length.

An intensification of the process in such cases
is possible only by intensifying the heat removal. The
simplest way to achieve this is to increase the effective
heat conductance λ of the contact mass with the aid of
suitable carriers.

Still greater intensification should be attainable by means of special constructions of contact apparatus
with intense external heat exchange, for example with the
catalyst supported on surface the other side of which is
cooled by some heat carrier. This type of apparatus permits sufficient lowering of the value of ξ, i.e., of the
temperature rise attained in the diffusional region, so
that even processes which do not tolerate a large temperature rise, can be carried out in the upper temperature
range.

This type of apparatus has not, so far, been

applied in present-day industrial practice, probably on
account of the necessary very thorough heat calculation
for which no method has as yet been developed.

Discussion of the problem of the temperature rise
of contact apparatus can be found in the work of Damköhler[11*]
However, this author is not aware even of the fundamental
fact that a stationary thermal regime, corresponding to
the reaction taking its course in the kinetic region, is
not always possible. He merely assumes the temperature
distribution over the cross-section to be parabolic and
calculates the maximum temperature rise in the center.
With a small temperature rise prescribed, one finds the
requirement of small dimensions of the apparatus, or of
high heat conductance of the contact mass, analogous to
ours. However, the author remains wholly unaware of the
fact that, when the dimensions of the apparatus are increased, the regime must break down with a discontinuous
passage to very high temperature rise in the diffusional
range, although this is exactly what is observed in practice.

Combustion of Coal

The foregoing general considerations can be
applied directly to the most important case of heterogeneous combustion, the combustion of coal.

It is true that direct application of our condition of ignition of a solid surface to the ignition of
coal can be complicated by the participation of volatile
matter and of CO in the ignition process. These compli-
cating factors are essential for the inflammation of coal
dust where the relative velocity of the gases with respect
to the coal particles is not great, and the particles move
surrounded by a jacket of gaseous products. In the instance
of combustion of a motionless particle in a flowing gas
these complications are less essential, the greater is the

---

[*] See also: Baron, T., Chem. Eng. Prog., 48 118 (1952).

velocity of the gas flow which immediately removes the
gaseous products from the particle. At the moment of the
ignition, the reaction $C + O_2$ is still in the kinetic
region; the reactions of evolution of volatile matter is
hardly sufficiently faster to appear in the diffusional
region and, if it remains in the kinetic region, this
means that no significant concentration of gaseous products
can accumulate at the surface. Our theory refers essentially to a high velocity of the gas flow with respect to
the solid surface; on this condition, it is probably fully
applicable to the combustion of coal.

Grozdovskii and Chukhanov[12] have established
that the process of combustion of coal proceeds different-
ly under different conditions, and gives different products. At low temperatures of the surrounding medium, low
concentrations of oxygen, and low velocities of the gas
flow, one observes a process producing a mixture of CO
and CO_2, with a strongly temperature dependent rate. At
high concentrations of oxygen, high velocities of the gas
flow, and high temperatures, the rate of the observed
process is practically independent of the temperature of
the surrounding medium and is proportional to the gas flow
velocity; the CO content in the product increases sharply.

Chukhanov[12] assumed that these two processes
represent two different reactions between carbon and oxygen; he called the first, oxidation, and the second, combustion. Actually[1] the two processes observed by Chukhanov
and Grozdovskii are not two different reactions, but two
different thermal regimes of the course of the same reaction. The experimental data of Chukhanov and his collaborators[13] are in qualitative agreement with our results.

As was shown by these investigators, the transition from the oxidation to the combustion process is
accomplished under conditions close to the conditions of
ignition of coal. Actually, this condition is discontinuous. There is no continuous transition from the oxidation

to the combustion regime. This at once lifts the apparent
conflict between the fact of the predominance of the combustion process at higher temperatures, and the practical
temperature-independence of its rate.

The sharp increase of the   CO   content in the
gaseous products on passing from the oxidation to the combustion range, is plausibly explained by the discontinuous
local temperature rise at the surface, as has been assumed
long ago by Lang[14]. In the combustion regime, the   CO
content in the gases should increase exponentially with
the surface temperature; under conditions where loss of
heat through radiation is essential, the surface temperature increases with the velocity of the gas flow, which
accounts for the phenomenon of high-velocity gasification.
Obviously a local temperature rise at the surface of the
coal cannot be detected by a thermocouple placed in the layer. Such a thermocouple, being in thermal equilibrium with
the surrounding medium, shows rather a temperature closer
to that of the gas flow, or a mean of the temperatures of
the gas and the surface. The only way to determine directly a local temperature rise would be to measure the surface temperature of the coal by means of an optical pyro-
meter. So far, such measurements were made only for isolated spherical particles of coal at low flow velocities
and, obviously, gave comparatively small values of the
temperature rise. Nobody who has visually observed a burning layer of coal can doubt the existence of local temperature rise in a layer of coal through which air is blown
at a high velocity.

The strong dependence of the rate of combustion
on the velocity of the gas flow and the concentration of
oxygen, observed by Chukhanov and Karzhavina[13] may be due
to an increase of the surface temperature with increasing
concentration of oxygen and increasing flow velocity under
conditions where loss of heat by radiation is essential.

We have, jointly with Klibanova[18], determined

the critical conditions of ignition of carbon filaments
in a flow of oxygen or air. The investigation of the
thermal regime and the critical conditions was applied as
a method of study of the kinetics of the reaction of carbon and oxygen under atmospheric pressure. The dependence
of the ignition temperature on the flow velocity gave the
activation energy, and its dependence of the oxygen concentration, the order of the reaction, which proved to be
distinctly lower than first. The latter conclusion is
further confirmed by the fact that critical conditions are
observed at an oxygen concentration of 2.5% and disappear
only at 0.8%, which corresponds to a value of the parameter ξ = 2.67. As was shown by Buben[8], critical conditions ought to be observed at values of ξ not lower
than $(1 + \sqrt{m})^2$ where m is the order of the reaction;
thus, our minimum value of ξ corresponds to a fractional
order of the reaction (for a first-order reaction, the
minimum value of ξ is, as was shown above, equal to 4).

Catalytic Oxidation of Isopropyl Alcohol

Of the greatest interest, from the point of view
of the theory of the thermal regime of heterogeneous exothermal reactions, are processes of strongly exothermic
catalysis. These processes are usually carried out in the
upper stationary thermal range, and involve high temperature differences between the gas and the catalyst surface.
The process is autothermal, i.e., needs no external heating. There can be critical phenomena of ignition and extinction of the surface. Owing to the latter fact, regulation of the surface temperature by way of modifying the
conditions of heat exchange, is far from always successful.

If one attempts to lower the surface temperature
too much, it can result in its extinction, and the process
will stop altogether. There exists a definite range of
unstable thermal regimes; the corresponding catalyst surface temperatures cannot be practically realized. The

higher the activation energy is, the narrower is the range
of stable and the broader the range of unstable thermal
regimes, and the smaller the chance of thermal regulation
of the process.

These considerations, based on the general theory
developed in the foregoing, proved useful for the technical process of catalytic oxidation of isopropyl alcohol to
acetone[9].

In inorganic chemical technology, the typical
instance of an exothermic catalytic process is the oxidation of ammonia to nitric acid; its counterpart in organic
technology are the processes of catalytic oxidation of
alcohols to aldehydes and ketones. The simplest among
them, the oxidation of methanol to formaldehyde, received
widespread application long ago.

More recently, catalytic oxidation of isopropyl
alcohol to acetone has taken considerable practical importance. All these processes are usually carried out on
metallic catalysts, platinum in the case of ammonia, and
copper or silver for alcohols.

Catalytic oxidation of alcohols was first done
on copper catalysts. It involves a very different thermal
regime. There is a strong surface temperature rise which
results in the development of side reactions, such as
formation of acids, which impair the quality of the product,
and of acetaldehyde and complete oxidation products (carbon dioxide and water) which lower the yield. On the
other hand, a lowering of the temperature by intensifying
the heat exchange, is unrealizable. Although the process
is accompanied by a strong temperature rise, it neverthe-
less proved to be quite precarious and thermally unstable.

This at first glance paradoxical situation could
be easily explained with the aid of the above theory of
the thermal regime of heterogeneous exothermic reactions.
Clearly, on a copper catalyst, the reaction has a high

activation energy, and its absolute rate is not very great.
Therefore, the range of stable stationary thermal regimes
is limited by very high temperatures. Any attempt to lower the temperature inevitably results in extinction of the
surface. This phenomenon of extinction, predicted by our
theory, is the key to the explanation of all the difficulties inherent in the thermal regime of the process with
a copper catalyst.

In the oxidation of isopropyl alcohol on a copper
catalyst, a stationary autothermal course of the process
proved to be entirely impracticable. The process could be
run without external heating, with only the aid of the
peculiar phenomenon of migration of the reaction zone.
The thickness of the catalyst layer was made very great,
and the reaction zone did not stay at one place but moved
along the catalyst layer alternatively in one and in the
opposite direction. When the reaction zone has reached
the end of the catalyst layer, the direction of the alcohol air feed was reversed, and thus the direction of the
migration was changed. This procedure facilitates the
thermal regime of the process, as the idle part of the
catalyst layer acts as heat exchanger. But, for the same
reason, the catalyst heats up strongly.

All these difficulties, due to the thermal regime
of the process, could be overcome by substituting a silver
catalyst. Silver, being a more active catalyst, has a
lower activation energy and therefore permits more stable
regulation of the temperature within wider limits without
fear of extinction. Use of a silver catalyst permitted a
considerable lowering of the temperature, and the elimination of the temperature rise resulted in drastic suppression
of side reactions.

We cannot here deal with the numerous purely
chemical problems linked with the development of this
process[9] as they have no direct bearing on the subject of
this book.

Catalytic Gas Analyzers

Another important application of the theory of the thermal regime of heterogeneous exothermic reactions is the construction of catalytic gas analyzers in which the content of a gas in a mixture is determined by the temperature rise of the surface of the catalyst on which this gas enters a catalytic reaction.

From the theory it follows that it is desirable to make such a gas analyzer operate in the upper stationary thermal range where its indications should not depend either on the activity of the catalyst or on the velocity of the gas, which presents great practical advantages.

An instrument based on that principle was constructed by Bresler and Zinovev for the analysis of oxygen.

Literature

1.   FRANK-KAMENETSKII, Zhur. Tekh. Fiz. 9, 1457 (1939).

2.   FRANK-KAMENETSKII, Doklady Akad. Nauk SSSR 30, 729 (1941).

3.   BUBEN, Sbornik rabot po fizicheskoi Khimii (collected works on Physical Chemistry) (Suppl. vol. to Zhur. Fiz. Khim. 1946) p. 148, 154 (1947).

4.   KARZHAVIN, Raschety po teknologii Svyazannogo azota (Calculations on the technology of fixed nitrogen) p. 104, Moscow 1935.

5.   ACKERMAN, VDI Forschungsheft No. 382, 1 (1937).

6.   BLINOV, Trudy Voronezh. Univ. 11, No. 1 (1939).

7.   DAVIES, Phil. Mag. 17, 233 (1934); 19, 309 (1935); 23, 409 (1937).

8.   BUBEN, Zhur. Fiz. Khim. 19, 250 (1945).

9.   FRANK-KAMENETSKII, Krentsel and Zverev, Khim. Promyshl. No. 1, 31 (1946).

10. BRESLER and ZINOVEV, Byull. Glavkisloroda No. 2, 24 (1945).

11. DAMKOHLER, in Eucken-Jacob's Chemie-Ingenieur, III, Teil 1, 448 (1937).

12. GROZDOVSKII and CHUKHANOV, Zhur. Priklad. Khim. No. 8, 1398 (1934); No. 1, 73 (1936); Khim. tverd. Topliva No. 9, 902; No. 10, 986 (1936).

13. CHUKHANOV, in "Protsess Goreniya uglya" (Combustion of coal), ed. Predvoditelev, p. 8 (1938).

14. LANG, Z. physik. Chem. 2, 181 (1888).

15. ZELDOVICH, Zhur. Eksptl. Teoret. Fiz. 11, 493, 501 (1941).

16. TODES and MARGOLIS, Izv. Akad. Nauk S.S.S.R., Otdel. Khim Nauk No. 1, 47; No. 3, 275 (1946).

17. MAYERS, Trans. ASME 59, 279 (1937); Ind. Eng. Chem. 32, 563 (1940).

18. KLIBANOVA and FRANK-KAMENETSKII, Acta physicochim. 18, 387 (1943).

CHAPTER X: PERIODIC PROCESSES IN CHEMICAL KINETICS

In the course of complex chemical reactions there are cases when the process becomes periodic. The heat evolved in the reaction can play a very important role. Therefore, the problem of periodic chemical reactions is intimately linked with macroscopic kinetics and particularly with the theory of combustion.

Conditions for an Oscillating Course of the Reaction in the Neighborhood of Quasistationary Concentrations

Let us consider the general problem of the course of a complex chemical process, using the convenient system of notation which we have introduced[1]. We designate the concentrations of the different intermediate products taking part in the process by x_i; the heat can also be treated as one of these products, with the temperature playing the role of its concentration.

The kinetics of the reaction is given by the system of kinetic equations

$$\frac{dx_i}{dt} = F_i(x_k) \qquad (X, 1)$$

where x_k stands for the entire set of x_k.

Equating the right-hand members of these equations to zero, we get a system of algebraic equations for
the determination of the stationary (or quasistationary)
values of x. If this system has finite positive real
solutions, we shall term these values X the quasistationary concentrations. In this case, the concentrations x
should either tend to the quasistationary values X or
oscillate around them, if the conditions derived in our
work[1] are fulfilled.

By quasistationary we mean the concentrations of
the intermediate products which can be expressed by the
concentrations of the initial substances but do not ex-
plicitly depend on time. The quasistationary concentrations do change with time, but only as a result of the
change of the concentrations of the initial substances.

Usually, in chemical kinetics, one uses the so-
called method of quasistationary concentrations in which
it is assumed that the concentrations x tend to the
quasistationary values X and attain them rapidly.

We shall find the conditions for the second alternative, namely an oscillating behavior of the process.
We designate the deviations of the concentrations x_1
from the quasistationary X_1 by ϵ_1

$$\epsilon_1 = x_1 - X_1 \qquad\qquad (X, 2)$$

Considering the magnitude ϵ small, we can put the system
of kinetic equations (X, 1) in the approximate form

$$\frac{d\epsilon_1}{dt} = \sum_k \left(\frac{\partial F_1}{\partial x_k}\right)_{x=X} \qquad \epsilon_k = \sum_k f_{1k}\epsilon_k \qquad (X, 3)$$

with the designation

$$f_{1k} = \left(\frac{\partial F_1}{\partial x_k}\right)_{x=X} \qquad\qquad (X, 4)$$

The magnitudes f_{1k} determine the character of the time
course of the process. They are therefore reasonably
termed the kinetic coefficients, and their matrix, the
kinetic matrix of the reaction.

The system of linear differential equations with
the constant coefficients (X, 3) has solutions of the form

$$\epsilon_i = \sum_k C_k e^{\mu_k t} \qquad (X, 5)$$

where μ_k are the roots of the characteristic equation

$$\begin{vmatrix} f_{11} - \mu & f_{12} & f_{13} & \cdots \\ f_{21} & f_{22} - \mu & f_{23} & \cdots \\ f_{31} & f_{32} & f_{33} - \mu & \cdots \\ \cdots & \cdots & \cdots & \cdots \end{vmatrix} = 0 \qquad (X, 6)$$

Consequently, the time course of the process is determined
by the properties of the roots of the characteristic equation μ_k.

If these roots are real, the process is not oscillating. If they have an imaginary part, the process is
oscillating. If the real part is negative, the oscillations will be damped (the equilibrium position stable).
If the real part is positive, the position of the equilibrium will be unstable, and the amplitude of the oscillation should increase with time.

In the latter case, the amplitude of the oscillations will either increase without limit (until the store
of initial reactant is exhausted) or it will tend to a
limiting value independent of the initial conditions
(limiting cycle). Such an undamped oscillation, with an
amplitude independent of the initial conditions, is termed
stationary self-exciting oscillation, or auto-oscillation.

Let us consider the simplest case when the system (X, 1) consists only of two equations. The character-

characteristic equation (X, 6) will take the form

$$\mu^2 - (f_{11} + f_{22})\mu + f_{11}f_{22} - f_{12}f_{21} = 0 \qquad (X, 7)$$

and its roots will be

$$\mu = \frac{f_{11} + f_{22}}{2} \pm \sqrt{\frac{(f_{11} + f_{22})^2}{4} - f_{11}f_{22} + f_{12}f_{21}} \quad (X, 8)$$

The latter expression can be rewritten in the form

$$\mu = \frac{f_{11} + f_{22}}{2} \pm \sqrt{\frac{(f_{11} - f_{22})^2}{4} + f_{12}f_{21}} \qquad (X, 9)$$

This shows immediately that an oscillatory course of the reaction is possible only when the magnitudes f_{12} and f_{21} have different signs and when

$$\sqrt{(-f_{12}f_{21})} > \left| \frac{f_{11} - f_{22}}{2} \right| \qquad (X, 10)$$

This is the general condition for an oscillatory course (in a sufficiently small range of quasistationary concentrations) of a chemical process the kinetics of which is described by a system of two kinetic equations.

From the point of view of mechanism, periodic chemical processes can be divided into purely kinetic oscillations linked only with a change of the concentrations of the intermediate reaction products, and thermokinetic oscillations in which, along with a change of the concentrations, periodic change of the temperature, due to the heat evolved in the reaction, also plays a substantial role.

Relaxation and Kinetic Oscillations

In addition to true kinetic oscillations, there can be also relaxation oscillations the frequency of which is determined by the rate of feed of the reacting mixture

into the reaction vessel. Such phenomena were observed,
for example, by Rayleigh[2] in the oxidation of vapors of
phosphorus, by Tokarev and Nekrasov[3] in the oxidation of
hydrogen. They can be observed in all cases where there
is a lower explosion limit. If, at the moment of introduction of the mixture into the vessel, its partial
pressure attains the lower limit, a flash will occur. If,
in this flash, all the mixture or a substantial part thereof is burned off (which will always be the case if the
flash is accompanied by a significant temperature rise),
the subsequently admitted mixture will fail to react until its pressure attains the lower limit, etc. We are not
here interested in such purely relaxation oscillations,
but will deal only with truly kinetic oscillations, the
frequency of which is linked with the kinetics of the
reaction.

The System of Volterra-Lotka

The simplest kinetic scheme leading to a periodic course of the reaction without participation of
thermal factors can be obtained on the assumption of autocatalysis by two consecutive intermediate products X
and Y.

Let the reaction proceed according to the scheme

$$A \longrightarrow X \longrightarrow Y \longrightarrow B$$

wherein the product X is formed autocatalytically from
the initial reactant A, and the product Y, also autocatalytically, from the product X. Designating the concentrations of A, X, Y, by a, x, y, we get the system
of kinetic equations

$$\frac{dx}{dt} = k_1 ax - k_2 xy$$

$$\frac{dy}{dt} = k_2 xy - k_3 ay$$

(X, 11)

where k_1, k_2, and k_3, are the rate constant of the individual reactions.

The quasistationary concentrations are

$$X = \frac{k_3 a}{k_2} \qquad\qquad (X, 12)$$

$$Y = \frac{k_1 a}{k_2} \qquad\qquad (X, 13)$$

For the magnitudes f_{1k} we get, according to (X, 4)

$$f_{11} = k_1 a - k_2 Y = 0 \qquad\qquad (X, 14)$$

$$f_{12} = - k_2 X = - k_3 a \qquad\qquad (X, 15)$$

$$f_{21} = k_2 Y = k_1 a \qquad\qquad (X, 16)$$

$$f_{22} = k_2 X - k_3 a = 0 \qquad\qquad (X, 17)$$

We see that condition (X, 10) is fulfilled identically for any values of the constants. Thus, the scheme (X, 11) describes a process which takes a periodic course at any values of the parameters. The frequency for small oscillations can be found from (X, 9), (X, 15), and (X, 16), and will be expressed by

$$v = \frac{a}{2\pi} \sqrt{k_1 k_3} \qquad\qquad (X, 18)$$

A system of the form (X, 11) was first considered by Lotka[4] and Volterra[5] in the mathematical theory of the struggle for life in which the role of X and Y was played by two kinds of animals devouring each other. Lotka[6] was the first the point out that equations of this form could play a role also in chemical kinetics.

For the explanation of cool-flame and two-stage ignition phenomena, we[7,8] have proposed a mechanism of

oxidation of higher hydrocarbons which leads to a system
of kinetic equations of the form (X, 11). We assumed that
oxidation of higher hydrocarbons, for example of gasoline,
proceeds over autocatalysis by two consecutive intermediate
products. The role of the first product X was attri-
buted to organic peroxides, and the role of the second
product Y, to aldehydes.

Periodic Processes in the Oxidation
of Hydrocarbons

        Our conclusion concerning the possibility of a
periodic course of the process is borne out by a number
of experimental facts. Thus, Newitt and Thornes[9],
Gerber and Pishchik in the laboratory of Neiman[10], Day
and Pease[11], and other investigators, observed a series
of consecutive cool-flame flashes when a mixture of pro-
pane and oxygen was admitted into a closed vessel pre-
heated to a certain temperature. However, only a limited
number of flashes can be observed under these conditions,
as the process soon comes to a halt as a result of exhaustion of the initial reactant.

        In our work with Gervart[12], we studied the peri-
odic pulsations of the cool flame under conditions of continuous feeding of a mixture of gasoline vapors with air
or oxygen into the vessel ("turbulent reactor") ensuring
complete mixing of the feed and the reacting mixture. Under
these conditions, it was possible to observe a stationary
periodic process over a long time. Flashes of the cool
flame occurred at a definite frequency, i.e., at regular
time intervals.

        Some of the quantitative laws observed are in
good agreement with our kinetic scheme which leads to the
system (X, 11) of kinetic equations. Thus, the frequency
of pulsations increases regularly with the concentration
of oxygen and with the temperature, and the temperature
dependence can be described by Arrhenius' law with an

activation energy of 15500 cal/mole. On the other hand,
the pulsation frequency was found to be independent of
total pressure and to decrease with increasing dimensions
of the vessel. These facts are not explainable on the
basis of this kinetic scheme.

The system (X, 11) has one more major shortcoming from the point of view of the theory of oscillations. The process described by this system represents
an oscillation, the amplitude of which is determined by
the initial conditions; in this respect, it is analogous
to a simple harmonic oscillation, for example to the motion
of an ideal pendulum without friction. Undamped oscillations of this kind, around a center-type equilibrium position, represent a unique case. The nature of the solution is changed qualitatively as soon as any additional
terms, no matter how small, are introduced into the equations; the oscillations either become damped or the amplitude increases. Inasmuch as, physically, the amplitude
cannot increase infinitely, we can expect auto-oscillations,
i.e., stable oscillations with an amplitude independent
of the initial conditions.

Our analysis, jointly with Salnikov[13], of the
influence of small additional terms on the character of
the solutions of the system (X, 11) showed that all corrections proportional to zero, and first powers of x and
y lead to damping, and only positive quadratic terms can
lead to oscillations with an increasing amplitude, and
under definite conditions, to a limiting cycle (auto-
oscillations).

Inasmuch as, physically, side processes not
taken into account in the scheme (X, 11) must always be
present, at least to a small extent (for example, the
generation of the intermediate product, not included in
the scheme (X, 11) according to which the reaction could
never start at x = 0), the real scheme should lead to
damped oscillations, unless a positive quadratic

autocatalysis (chain interaction in Semenov's[14] terminology)
is also introduced. Thus, it appears difficult to account
for a prolonged stable periodic course of the process with
the aid of the scheme (X, 11).

Gorelik was the first to point out that this
difficulty could be surmounted by taking into account the
heat evolved in the process. It is quite probable that
small additional corrections of a thermal nature could
convert the process described by the system (X, 11) into
an autooscillatory one. But this idea of heat-stabilized
chemical oscillations has not so far received an adequate
mathematical formulation.

Thermokinetic Oscillations

Besides purely kinetic oscillations, there is
the possibility of a periodic course of a chemical reaction due at the same time to both the kinetics of the reaction and to the evolution and removal of the heat. We
shall term this kind of oscillation thermokinetic.

The simplest form of thermokinetic oscillations
is possible when one intermediate product X takes part
in the reaction. Let the reaction take place according
to the scheme

$$A \longrightarrow X \longrightarrow B$$

and let the rate of the second stage be more strongly
temperature dependent. The complete description of the
process is given by the system of two equations, one
kinetic and the other thermal

$$\frac{dx}{dt} = f - k(T)x$$

$$\frac{dT}{dt} = \frac{q}{c\rho} - \frac{\alpha S}{c\rho\omega} (T - T_0)$$

(X, 19)

Here f is the rate of formation of the intermediate
product X from the initial reactant A; $k(T)$ is the
rate constant of the conversion of X into the final
product B, which should depend on the temperature more
strongly that f; q is the rate of evolution of heat,
which depends on the rates and the thermal effects of both
reactions; α is the heat exchange coefficient; T is
the temperature of the reacting mixture; T_0 is the
temperature of the surrounding medium; c is the heat
capacity of the reacting mixture, ρ is its density, ω
is the volume of the reacting system, S is the outer surface through which the heat is exchanged.

Provided the rate of the second reaction depends
on the temperature more strongly than the rate of the
second reaction, and heat evolution is linked primarily
with the second reaction, the process can become oscillatory.
Accumulation of the product X will lead to enhanced heat
evolution as a result of the second reaction and of the
temperature rise. A rise of the temperature will increase
the rate of consumption of the product X more strongly
than the rate of its formation. The concentration of the
product X will decrease, which will lead to a lowering
of the temperature, and so forth.

This conclusion is also corroborated by mathematical analysis. With different concrete assumptions about
the temperature and concentration dependence of f, k,
and q, one can demonstrate the possibility of oscillations
and, under definite conditions, of auto-oscillations.

Thermokinetic oscillations can play a role in
two fields: in ordinary chemical kinetics and in the theory
of the internal structure of stars.

In ordinary chemical kinetics, rates of reaction
depend on the temperature according to Arrhenius' law. In
this case, the magnitudes f and k can be represented
in the form

X. PERIODIC PROCESSES

$$f = z_1 e^{-\frac{E_1}{RT}}; \quad k = z_2 e^{-\frac{E_2}{RT}} \qquad (X, 20)$$

For thermokinetic oscillations to be possible, it is necessary that

$$E_2 > E_1$$

If both the first and the second reaction have a thermal effect, the rate of evolution of heat will be expressed as

$$q = Q_1 z_1 e^{-\frac{E_1}{RT}} + Q_2 z_2 e^{-\frac{E_2}{RT}} \qquad (X, 21)$$

The system (X, 19) with the expressions (X, 20) and (X, 21) for f, k, and q (with $Q_1 = 0$) was investigated by Salnikov who showed that its solutions can be oscillatory and, under certain conditions, also periodic, i.e., auto-oscillatory. Further development of that work ought to show whether it might not be possible to interpret the above-mentioned periodic phenomena in the oxidation of hydrocarbons as thermokinetic oscillations.

In the theory of the internal structure of stars, the role of Arrhenius' law, which differs from it only in that $\sqrt[3]{T}$ stands instead of the temperature.

According to Bethe[15], two kinds of nuclear reactions can be the source of the energy of the stars: reactions between light nuclei (direct addition of two protons and the immediately following reactions) and reactions involving heavy nuclei (Bethe's carbon cycle); the temperature dependence of the reactions of the first group is much weaker than for the second group.

Bethe considers the concentration of heavy nuclei in a star constant. If, however, one admits that this assumption is insufficiently substantiated, and grants the possibility of hypothetic processes of formation and further transformation of heavy nuclei in stars (sufficiently probable instances of such processes can be quoted), one can assume in stars thermokinetic oscillations of the type described by the system (X, 19) with kinetics corresponding to Atkinson's law.

If so, it is possible that periodic processes (including processes with very long periods, exceeding by far possible times of human observation) play a much greater role in the life of stars than is commonly assumed.

Literature

1. FRANK-KAMENETSKII Zhur. Fiz. Khim. 14, 695 (1940).

2. RAYLEIGH, Proc. Roy. Soc. A 99, 372 (1921).

3. TOKAREV and NEKRASOV, Zhur. Fiz. Khim. 8, 504 (1936).

4. LOTKA, Elements of physical biology. Baltimore, 1925.

5. VOLTERRA, Leçons sur la théorie mathematique de la lutte pour la vie. Paris, 1932.

6. LOTKA, J. Am. Chem. Soc. 42, 1595 (1920).

7. FRANK-KAMENETSKII, Doklady Akad. Nauk S.S.S.R. 25, 672 (1939).

8. FRANK-KAMENETSKII, Zhur. Fiz. Khim. 14, 30 (1940).

9. NEWITT and THORNES, J. Chem. Soc. 1937, 1669.

10. PISHCHIK, Thesis, Leningrad Ind. Inst. 1939.

11. DAY and PEASE, J. Am. Chem. Soc. 62, 2234 (1940).

12. GERVART and FRANK-KAMENETSKII, Izvest. Akad. Nauk S.S.S.R., Otdel. Khim. Nauk No. 4, p. 210 (1942).

13. FRANK-KAMENETSKII and SALNIKOV, Zhur. Fiz. Khim. 17, 79 (1943).

14. SEMENOV, Tsepnye reaktsii (Chain reactions) 1934.

15. BETHE, Phys. Rev. 55, 434 (1939).

NOTATION

A	turbulent exchange coefficient	
A	proportionality factor in formula	(II, 68)
A	maximum work of a chemical reaction	
a	temperature conductivity coefficient	
a	auxiliary constant	(V, 33)
a	integration constant	(VII, 10)
a	kinetic constant in the theory of second-order autocatalysis	(VIII, 29)
B	dimensionless maximum temperature	(VI, 25)
B	integration constant	
b	integration constant	(VII, 10)
b	constant in formula	(IV, 3)
C	concentration	
ΔC	concentration difference	
C	resistance coefficient (in external flow)	
C	Euler constant	
c	heat capacity	
cosh	hyperbolic cosine	
D	diffusion coefficient	
d	determining linear dimension	
div	symbol of divergence	
E	activation energy	
E_1	integral logarithm	
e	base of natural logarithms	
F	dimensionless factor in the theory of flame propagation	(VIII, 24)
f	resistance coefficient (over a length equal to the hydraulic radius)	
f	rate constant of the chain branching reaction	
$f(\eta) = f\left(\frac{\gamma}{y}\right)$	linking function	(V, 76)
f_{1k}	kinetic coefficients	(X, 4)

Equation
Number

G	constant in the theory of the laminar sublayer	(V, 89)
grad	symbol of the gradient	
g	rate constant of the chain termination reaction	
g	gravity acceleration	
I	heat content	
J	Bessel function	
J_o	zero-order Bessel function	
k_T	thermodiffusion ratio	(IV, 1)
k	reaction rate constant	
k^*	effective rate constant	(II, 7)
L	dimensionless thickness of the laminar sublayer	(V, 6)
L	length	
l	mixing length	
ln	natural logarithm	
log	decadic logarithm	
M	Margoulis criterion	
M_∞	limiting value of the Margoulis criterion at large values of the Prandtl criterion	(V, 7)
m	order of reaction	(VII, 26)
m	exponent of Reynolds criterion	(I, 41)
Nu	Nusselt criterion	
Nu_D	diffusional Nusselt criterion	
Nu_λ	thermal Nusselt criterion	
n	exponent of the Prandtl criterion in the law	(I, 41)
n	exponent in the expression of the dependence of the turbulent exchange coefficient on the distance	(V, 24)
n	order of reaction	
n_o	rate of chain generation	
n	relative reactant concentration	(VI, 37)
n	integral number	
P	total pressure	

p	partial pressure	
p^o	partial pressure at a distance from the surface	
p_s	saturated vapor pressure	
Δp	partial pressure difference	
Pe	Péclet criterion	
Pr	Prandtl criterion	
Q	heat effect of a reaction	
$Q(x)$	function expressing the temperature or concentration dependence of the chemical reaction rate	(VIII, 2)
q'	density of sources of heat or substance	
q	heat or diffusion flow	
Re	Reynolds criterion	
R	gas constant	
r'	hydraulic radius	
r	radius of vessel	
S	surface area	
T	absolute temperature	
ΔT	temperature difference	
T_m^*	theoretical maximum temperature of the reaction, calculated on the assumption of constant heat capacity	(VI, 30)
T_m	maximum combustion temperature	
t	time	
u	mean velocity of molecular motion (in the Maxwell-Stephan temperature)	
u	pulsation velocity	
u_o	value of the pulsation velocity at a distance from the wall	
u^*	dimensionless pulsation velocity	
u	dimensionless temperature	(VI, 6)
u_x	longitudinal components of the velocity	
u_y	transverse components of the velocity	

Equation
Number

v	velocity of flow	
v'	value of the velocity at the boundary of the laminar sublayer	
v_x	longitudinal components of the velocity	
v_y	transverse components of the velocity	
$v*$	dimensionless velocity	(V, 96)
v_x	mean relative velocity of motion of two particles of gas or liquid, at a distance x from each other	(V, 102)
w	velocity of flame propagation	
X_i	quasistationary concentrations of intermediate products	(X, 2)
X	function depending only on the coordinates in Fourier's method	
x	space coordinate, or totality of all the space coordinates	
x	mole fraction	
x_j, x_k	running coordinates of the intermediate products	
x	auxiliary variable	
y	distance from the surface	
y_o	value of y at the point where the characteristic values of ΔC, ΔT, V, are reached	
y	derivative of temperature with respect to the coordinate	(VI, 31)
$y = \dfrac{dx}{d\xi}$	auxiliary variable in the theory of flame propagation	(VIII, 3)
$y*$	dimensionless distance from the wall	(V, 96)
Z	dimensionless magnitude in the theory of condensation of vapors in the presence of incondensible gases	
z	variable distance over the length of the tube, in the calculation of the local value of the Nusselt criterion	
z	diameter of a disperse particle	

z	pre-exponential factor in Arrhenius' law	
α	heat exchange coefficient	
β	diffusion velocity constant	
β_∞	limiting value of the diffusion velocity constant at large values of the Prandtl criterion	(V, 7)
γ	width of the linking zone	(V, 76)
Δ	Laplace operator	
Δ_1	Laplace operator in dimensionless coordinates	(VI, 13)
ΔT	temperature difference	
ΔC	concentration difference	
Δp	partial pressure difference	
δ	thickness of the reduced film	
δ_D	the same for diffusion	
δ_λ	the same for heat transfer	
δ'	thickness of the boundary layer	(V, 1)
$\delta*$	thickness of the laminar sublayer	
$\delta*$	approximate value of the parameter δ	(VII, 24)
δD	thickness of the diffusion layer after Tunitskii	(V, 108)
ϵ	dissociation energy	(V, 102)
ϵ	shift of the ignition limit as a result of combustion	(VII, 26)
ϵ_1	deviation of the concentration x_1 from the quasistationary value X_1	(X, 2)
ζ	dimensionless concentration	
ζ_1	concentration within a layer of porous material	
ζ'	dimensionless concentration within a layer of porous material	
ζ	dimensionless parameter in the theory of the thermal regimen of a layer or channel	(IX, 59)
ζ	resistance coefficient (over a length equal to the diameter)	

η	auxiliary variable in the theory of the laminar sublayer	(V, 77)
	temperature difference	(VI, 12)
θ	dimensionless temperature	(VI, 12)
θ_m	dimensionless maximum temperature of combustion	(VI, 47)
κ	constant factor in the expression for the turbulent exchange coefficient	(V, 24) (V, 67)
λ	heat conductivity coefficient	
μ_k	roots of the zero-order Bessel function	
$\mu_1 = 2.4048$	first root of the zero-order Bessel function	
μ	dimensionless parameter in the diffusional kinetics of fractional-order reactions	
μ	dimensionless parameter in the theory of thermal ignition for heterogeneous reactions	(IX, 8)
μ	dimensionless flame propagation velocity	(VIII, 4) (VIII, 5)
μ_k	roots of the characteristic equation	(X, 10)
μ	dynamic viscosity	
ν	kinematic viscosity	
ν	dimensionless parameter	(VI, 8) (IX, 59)
ξ	dimensionless coordinate	
ξ	coordinate in a system bound with the flame front	
ξ	dimensionless parameter in the theory of thermal ignition for heterogeneous reactions	(IX, 11)
ξ_1	thermal thickness of the flame front(VIII, 21)	(VIII, 21)
ξ_2	chemical thickness of the flame front	(VIII, 23)
Π	perimeter	
ρ	density	
σ	surface area	

		Equation Number
σ	auxiliary magnitude	(VII, 13) (IX, 45)
τ	characteristic time	
τ	auxiliary magnitude	(IX, 46)
τ_0	tangential stress at the surface	
φ	coefficient of self-acceleration of autocatalytic reactions	
φ	ratio of the velocity at the boundary of the laminar sub-layer and the mean velocity of the flow	(V, 37)
φ	auxiliary function	
ω	volume	
	function depending only on the time in Fourier's method